WITHDRAWN

Carnegie Mellon

Tropospheric Ozone Abatement

Springer
*Berlin
Heidelberg
New York
Barcelona
Hong Kong
London
Milan
Paris
Singapore
Tokyo*

Rainer Friedrich · Stefan Reis (Eds.)

Tropospheric Ozone Abatement

Developing Efficient Strategies
for the Reduction
of Ozone Precursor Emissions
in Europe

With 104 Figures
and 62 Tables

Springer

Prof. Dr. Rainer Friedrich
Dipl.-Wirtsch.-Ing. Stefan Reis
University of Stuttgart
Institute of Energy Economics and the Rational Use of Energy
Heßbrühlstraße 49a
D-70550 Stuttgart, Germany

ISBN 3-540-66614-1 Springer-Verlag Berlin Heidelberg New York

Library of Congress Cataloging-in-Publication Data applied for
Die Deutsche Bibliothek – CIP-Einheitsaufnahme
Tropospheric ozone abatement: developing efficient strategies for the reduction of ozone precursor emissions in Europe; with 62 tables / Rainer Friedrich; Stefan Reis (ed.). – Berlin; Heidelberg; New York; Barcelona; Hong Kong; London; Milan; Paris; Singapore; Tokyo: Springer, 2000
ISBN 3-540-66614-1

This work is subject to copyright. All rights are reserved, whether the whole or part of the material is concerned, specifically the rights of translation, reprinting, reuse of illustrations, recitation, broadcasting, reproduction on microfilm or in any other way, and storage in data banks. Duplication of this publication or parts thereof is permitted only under the provisions of the German Copyright Law of September 9, 1965, in its current version, and permission for use must always be obtained from Springer-Verlag. Violations are liable for prosecution under the German Copyright Law.

© Springer-Verlag Berlin · Heidelberg 2000
Printed in Germany

The use of general descriptive names, registered names, trademarks, etc. in this publication does not imply, even in the absence of a specific statement, that such names are exempt from the relevant protective laws and regulations and therefore free for general use.

Cover-Design: de'blik, Berlin
Typesetting: camera-ready by the editors

SPIN 10735712   30/3136xz-5 4 3 2 1 0 – Printed on acid-free paper

Preface

The occurrence of high concentrations of ozone during summer episodes in the troposphere over Europe is a problem still unsolved. Although a number of measures have been implemented that will achieve a further reduction of precursor emissions in the next years, this will not be sufficient to reduce the ozone concentration to levels below thresholds set up to protect human health and plants. Thus, further reductions of emissions of volatile organic compounds and nitrogen oxides are necessary. However, with regard to the increasing costs associated with increasing emission reduction, it is essential to implement abatement strategies, that are effective, i.e. achieving the environmental aim set up, and efficient, i. e. doing this with the least costs possible.

In this book, the authors describe the features and the application of a methodology and a model system to identify effective and efficient strategies to reduce ambient concentrations of tropospheric ozone to comply with thresholds set up to protect human health, agricultural crops and ecosystems. Furthermore, macroeconomic impacts of such strategies are addressed and, as burden and benefits of these strategies are not equally distributed between countries, different burden sharing schemes are discussed.

The content of this book is based on results of a comprehensive research project, the project INFOS (assessment of policy instruments for efficient ozone abatement strategies in Europe), funded by the European Commission (Directorate General XII) under the Fourth Framework Programme for research, technological development and demonstration activities.

Rainer Friedrich, Stefan Reis

　　　　3.2.2. Per Capita Emissions..30
　　　　3.2.3 Time Series Analysis..32
　　3.3 References 33

4 Emission Abatement Measures, D. Papameletiou, J. M. Maqueda **35**
　　4.1 Introduction 35
　　4.2 Road Transport Sector...36
　　　　4.2.1 Passenger Cars and Light-Duty Vehicles – Gasoline..................41
　　　　4.2.2 Passenger Cars and Light-Duty Vehicles – Diesel......................42
　　　　4.2.3 Heavy-Duty Vehicles – Diesel..43
　　　　4.2.4 Fuel Cell Powered Vehicles 44
　　　　4.2.5 Techno–Economic Profiles of Measures for Road Transport 45
　　4.3 Energy Sector 46
　　　　4.3.1 Techno–Economic Profile of Measures for the Energy Sector ...48
　　4.4 Solvent Use Sector 51
　　　　4.4.1 Techno–Economic Profiles of Measures for the Solvent Sector.53
　　4.5 Conclusions 56
　　4.6 References 57
　　　　Acknowledgement..60

5 Scenarios of Future Development, S. Reis, R. Friedrich **61**
　　5.1 Methodology for Emission Projection...61
　　　　5.1.1 Approach...61
　　　　5.1.2 Emission Projection – Basic Methodology 61
　　5.2 Driving Forces of Emission Development 62
　　　　5.2.1 Societal and Demographic Trends 63
　　　　5.2.2 Economic Trends 63
　　　　5.2.3 Energy Trends 64
　　5.3 Political and Legislative Framework...66
　　　　5.3.1 Sectoral Air Pollution Control Policies 67
　　　　5.3.2 Legislation on Specific Substances 69
　　5.4 Detailed Sectoral Projection 71
　　　　5.4.1 Emissions from the Energy Sector 71
　　　　5.4.2 Emissions from Residential and Commercial Combustion.........72
　　　　5.4.3 Emissions from Industrial Combustion.......................................72
　　　　5.4.4 Emissions from Industry Processes...73
　　　　5.4.5 Emissions from Fuel Handling..73
　　　　5.4.6 Emissions from Solvent Use 74
　　　　5.4.7 Emissions from Road Transport..75
　　　　5.4.8 Emissions from Other Mobile Sources 76
　　　　5.4.9 Emissions from Waste Handling...77
　　　　5.4.10 Emissions from Agriculture 77
　　5.5 Trend Scenario Emissions 77
　　5.6 References 81

6 Regional Modeling of Tropospheric Ozone, D. Simpson, J. E. Jonson 83
- 6.1 Introduction 83
- 6.2 The Models..83
- 6.3 Emission Input to Models..84
- 6.4 Statistics Used 86
- 6.5 Model Calculations: 1990 Base and 2010 Trend Case 86
  - 6.5.1 Base-Case 1990 and Trend 2010 Scenario, 5 yr Average 87
  - 6.5.2 Comparison of Models: Trend 2010 Scenario.................87
- 6.6 Model Calculations: Sector-by-Sector Emission Scenarios...................88
  - 6.6.1 Emission Control Measures, SNAP sectors 1-10 88
  - 6.6.2 Road Traffic Measures 90
- 6.7 Country-to-Country Calculations 92
- 6.8 Summary and Conclusions 93
- 6.9 References 96

7 Optimising Regional Ozone Reduction Strategies, S. Reis, D. Simpson, R. Friedrich 99
- 7.1 Introduction 99
- 7.2 The OMEGA Model..100
  - 7.2.1 Inputs to the Optimisation 103
  - 7.2.2 An Optimised Solution?...108
- 7.3 OMEGA Model Results 109
- 7.4 Evaluation of Optimisation Results...116
- 7.5 Modelled Exceedance Statistics of Ozone Thresholds 118
- 7.6 Summary and Conclusions 119
- 7.7 References 119
  - Acknowledgements 120

8 Tropospheric Ozone and Urban Air Quality, N. Moussiopoulos, P. Sahm, P.-M. Tourlou, T. Nitis, A. K. Azad and S. Papalexiou 121
- 8.1 Introduction 121
- 8.2 Ambient Ozone Concentrations in Selected Mesoscale Areas 122
  - 8.2.1 Emission Estimates 123
  - 8.2.2 Ambient Concentrations of Tropospheric Ozone in Athens.......126
  - 8.2.3 Ambient Concentrations of Tropospheric Ozone in Stuttgart ...135
  - 8.2.4 Ambient Concentrations of Tropospheric Ozone in Milan 143
- 8.3 Conclusions 147
- 8.4 References 149

9 Efficiency, Equity and Burden-Sharing, R. Salmons 151
- 9.1 Introduction 151
- 9.2 Efficiency 152
  - 9.2.1 Permit Trading and Pollution Characteristics............................153

Abbreviations and Acronyms

ACEA	Association des Constructeurs Européens d'Automobiles
AOTxx	Accumulated Ozone Exposure over a Threshold of xx ppb
APP	Ability to Pay Principle
BAU	Business As Usual
CLRTAP	Convention on Longe Range Transboundary Air Pollution
CO	Carbon Monoxide
CORINAIR	Emission Inventory for Air Pollutants under CORINE
CORINE	Co-ordination d'Information Environmental
DG	Directorate General
EAP	Environmental Action Plan
EC	European Commission
ECE	*see* UNECE
EF	Emission Factor
EMEP	European Monitoring and Evaluation Programme
ETCAQ	European Topic Centre Air Quality
EU	European Union
EUDC	Extra Urban Driving Cycle
EUROPIA	European Petroleum Industry Association
EUROSTAT	Statistical Office of the European Communities
EUROTRAC	Project on the Transport and Chemical Transformation of Environmentally Relevant Trace Constituents in the Troposphere over Europe
EZM	European Zooming Model
GC	Gap Closure
GDP	Gross Domestic Product
GENEMIS	Generation of Emission Data for Episodes (EUROTRAC Subproject)
GIS	Geographical Information System
GNP	Gross National Product
HC	Hydrocarbons
HDV	Heavy Duty Vehicle
IPCC	Intergovernmental Panel on Climate Change
IPPC	Integrated Pollution Prevention and Control
LCP	Large Combustion Plant
LDV	Light Duty Vehicle
MC	Motorcycle
MP	Moped
MSC-W	Meteorological Synthesizing Centre - West
NEDC	New European Driving Cycle
(NM)VOC	(Non-Methane) Volatile Organic Compound
NO_x	Nitrogen Oxide
O&M	Operation and Maintenance
OECD	Organisation for Economic Co-operation and Development
PC	Passenger Car
PM	Particulate Matter
ppm (ppb)	parts per million (billion)

PPP	Polluter Pays Principle
SCR	Selective Catalytic Reduction
SNAP	Selected Nomenclature for Air Pollutants
SNCR	Selective Non-Catalytic Reduction
SO_2	Sulfur Dioxide
TWC	Three-Way Catalyst
UNECE	United Nations Economic Commission for Europe
UNFCCC	United Nations Framework Convention on Climate Change
US EPA	US Environmental Protection Agency
VPP	Victim Pays Principle
WHO	World Health Organization

List of Figures

Fig. 2.1. Tropospheric cycles of catalysis 6
Fig. 2.2. Sectoral contributions to emissions of NO_x for the EU15 in 1994 8
Fig. 2.3. Sectoral contributions to emissions of NMVOC for the EU15 in 1994 ... 8
Fig. 2.4. Spatial distribution of NO_x emissions in Europe 1994 9
Fig. 2.5. Spatial distribution of NMVOC emissions in Europe 1994 9
Fig. 2.6. Yield-loss versus AOT40 relationship 12
Fig. 2.7. Areas covered by the EC's Air Quality Framework Directive 17
Fig. 2.8. European network of measuring stations for tropospheric ozone 22

Fig. 3.1. Shares of fossil fuels for power generation in Europe 1990 27
Fig. 3.2. NO_x emissions of EU15 countries 1994 29
Fig. 3.3. NMVOC emissions of EU15 countries 1994 30
Fig. 3.4. Per capita emissions of NO_x in 1994 31
Fig. 3.5. Per capita emissions of NMVOC in 1994 31
Fig. 3.6. Time series of NO_x emissions in Europe 32
Fig. 3.7. Time series of NMVOC emissions in Europe 32

Fig. 4.1. Applied work approach and data model 35

Fig. 5.1. Population growth in the EU15 – 1990 vs. 2010 63
Fig. 5.2. Growth of GDP in the EU15 – 1990 vs. 2010 64
Fig. 5.3. Shares of fossil fuels used in electricity production in EU15 countries .. 64
Fig. 5.4. Changes in energy demand from solid fuels – 1990 vs. 2010 65
Fig. 5.5. Changes in energy demand from oil – 1990 vs. 2010 65
Fig. 5.6. Changes in energy demand from natural gas – 1990 vs. 2010 66
Fig. 5.7. Changes in fuel demand for the transport sector in the EU15 74
Fig. 5.8. Reductions of total EU15 emissions of NO_x and NMVOC 78
Fig. 5.9. 2010 NO_x trend emissions vs CORINAIR 90 emissions 78
Fig. 5.10. 2010 NMVOC trend emissions vs CORINAIR 90 emissions 79

Fig. 5.11. 2010 NO_x sectoral trend emissions vs. CORINAIR 90 emissions 80

Fig. 5.12. 2010 NMVOC sectoral trend emissions vs. CORINAIR 90 emissions. 81

Fig. 7.1. Sketch of iteration procedure... 101

Fig. 7.2. Illustrative ozone isopleth diagrams.. 102

Fig. 7.3. Exemplary abatement cost curve for NO_x 105

Fig. 7.4. Exemplary abatement cost curve for NMVOC 106

Fig. 7.5. Reduction of NO_x emissions necessary for each country to achieve given GC of AOT60 health 111

Fig. 7.6. Reduction of NMVOC emissions ecessary for each country to achieve given GC of AOT60 health 111

Fig. 7.7. Total abatement costs to achieve the given Gap Closures of AOT 60 .. 113

Fig. 7.8. Total abatement costs to achieve 33 % Gap Closure of AOT 60 vs. benefits due to avoided health damages 113

Fig. 7.9. Total abatement costs to achieve 33 % Gap Closure of AOT 40 vs. benefits due to avoided crop damages 114

Fig. 7.10. Reductions of total EU15 NO_x emissions necessary to achieve specified gap closures of AOT40crops, AOT40forests and AOT60 115

Fig. 7.11. Reductions of total EU15 NMVOC emissions necessary to achieve specified gap closures of AOT40crops, AOT40forests and AOT60 116

Fig. 7.12. Comparison of the absolute AOT40f values and the changes, $\Delta AOT40_f$, calculated by the OMEGA model and the full EMEP oxidant model 117

Fig. 7.13. Comparison of the absolute AOT60 values and the changes, $\Delta AOT60$, calculated by the OMEGA model and the full EMEP oxidant model. 117

Fig. 8.1. Ozone concentration downwind a major source (e.g. a city)................. 122

Fig. 8.2. Topography of the Greater Athens area 126

Fig. 8.3. Model results for the accumulated exposure to ozone for the Greater Athens area according to the reference case.. 129

Fig. 8.4. Calculated differences between the "business as usual" 2010 scenario and the reference case for the Greater Athens area 130

Fig. 8.5. Calculated differences between the "business as usual" 2010 scenario and the reference case and the additional 50% reduction in NO_x emissions on the local scale for the Greater Athens area........................ 132

Fig. 8.6. Calculated differences between the "business as usual" 2010 scenario and the reference case and the additional 50% reduction in VOC emissions on the local scale for the Greater Athens area................................... 133

Fig. 8.7. Days with the running 8h mean ozone concentration exceeding 120 µg/m^3 and wind direction statistics in the state of Baden-Wuerttemberg 136

Fig. 8.8. Model results for the accumulated exposure to ozone for the Greater Stuttgart area according to the reference case... 137

Fig. 8.9. Calculated differences between the business as usual 2010 trend scenario and the reference case for the Greater Stuttgart area............................ 139

Fig. 8.10. Calculated differences between the "business as usual" 2010 scenario and the reference caseand additional 50% reduction in NO_x emissions on the local scale for the Greater Stuttgart area 141

Fig. 8.11. Calculated differences between the "business as usual" 2010 scenario and the reference case and additional 50% reduction in VOC emissions on the local scale for the Greater Stuttgart area 142

Fig. 8.12. Model results for the wind field and ozone concentrations for the Milan metropolitan area according to the reference case 145

Fig. 8.13. Calculated differences of ozone concentrations between the "business as usual" 2010 scenario and the reference case for the Milan metropolitan area 145

Fig. 8.14. Calculated differences of ozone concentrations between the "business as usual" 2010 scenario with a further 50% reduction in local NO_x emissions and the reference case for the Milan metropolitan area 146

Fig. 8.15. Calculated differences of ozone concentrations between the "business as usual" 2010 scenario with a further 50% reduction in VOC emissions and the reference case for the Milan metropolitan area 146

Fig. 9.1. Global ozone control problem 155

Fig. 9.2. Local "linearized" ozone control problem.. 158

Fig. 9.3. Concept of iterative zonal permit trading 162

Fig. 9.4. Cost profiles over time for different revision bounds............................. 164

Fig. 9.5. Marginal abatement cost curves 170

Fig. 9.6. Comparison of estimated and actual cost minimum solutions............... 170

Fig. 9.7. Abatement costs as % of GDP.. 180

Fig. 9.8. Alternative interpretations of polluter pays principle 181

Fig. 9.9. Relationship between ozone concentration and NO_x emissions 184

Fig. 9.10. Transfer payments under alternative equity principles........................ 189

concentrations only. OMEGA was run with generic cost-curves to assess, how far emissions of ozone precursors would have to be reduced to achieve a specific gap closure of AOT 40 for crops

Plate 13. Ozone concentration plots for different gap closure attempts for AOT $40_{forests}$

Plate 14. Ozone concentration plots for an optimisation scenario for AOT 60 (health) calculated by the OMEGA Optimisation Model

Plate 15. Ozone concentration plots for an optimisation scenario for AOT 40_{crops} calculated by the OMEGA Optimisation Model

Plate 16. Ozone concentration plots for an optimisation scenario for AOT $40_{forests}$ calculated by OMEGA Optimisation Model

List of Tables

Table 2.1. Health impacts of different ambient concentrations of ground level ozone.. 11

Table 2.2. Protocols to the CLRTAP 15

Table 2.3. Critical Levels for Vegetation 16

Table 2.4. Limit Values of the Air Quality Framework Directive.......................... 18

Table 2.5. Ozone threshold values set by the EC Directive on Air Pollution by Ozone... 19

Table 2.6. Ozone target values and objectives set by the Draft Daughter Directive on Ozone.. 20

Table 3.1. Shares of sectoral emissions of NO_x in the EU15 1990....................... 25

Table 3.2. Shares of emissions of NMVOC in the EU15 1994 28

Table 4.1. Overview of Auto-Oil data coverage on advanced motor vehicle technologies and emission abatement techniques and data used in this study....... 38

Table 4.2. Advanced motor vehicle techniques selected as abatement options in this study 39

Table 4.3. Abatement options for the road transport sector.................................... 45

Table 4.4. Efficiency and Cost estimation of Automotive Fuel Cell Systems....... 46

Table 4.5. Fossil Fuelled Power Plants in Europe – Installed NO_x Abatement Capacities 49

Table 4.6. Technology measures for emission reduction in the energy production sector: Large Combustion Plants (LCP) and institutional and commercial combustion plants 50

Table 4.7. NMVOC emitting sectors and related activities 52

Table 4.8. Abatement measures for the solvent sector 54

Table 5.1. Overview on legislation concerning road transport vehicles in the EU (and related ECE regulations) 67

Table 5.2. Emission reductions achievable by EURO 1 – 4 standards related to pre-EURO levels 68

Table 5.3. Activities covered by the EC Solvent Directive 70

Table 5.4. NO_x emissions in the trend scenario for 2010 compared with CORINAIR emission inventory data (ktonnes) 79

Table 5.5. Anthropogenic NMVOC emissions in the trend scenario for 2010 compared with CORINAIR emission inventory data 80

Table 6.1. EMEP MSC-W Lagrangian and Eulerian models, brief summary. 85

Table 6.2. Emission projection (year 2010 base-case) and mean reductions in European AOT40 (forests) and AOT60. 89

Table 6.3. Emission projection (year 2010 base-case) and calculated reductions in mean EU-15 AOT40 and AOT60 90

Table 6.4. Calculated changes in percentiles of $AOT40_f$ for each emission scenario. Results from two models. 91

Table 6.5. Calculated changes in percentiles of AOT60 for each emission scenario. Results from two models 91

Table 6.6. Country-to-country matrix for 40 % NO_x Reduction, AOT40-crops 94

Table 6.7. Country-to-country matrix for 40 % VOC Reduction, AOT40-crops .. 94

Table 6.8. Country-to-country matrix for 40 % NO_x Reduction, AOT60 95

Table 6.9. Country-to-country matrix for 40 % VOC Reduction, AOT60 95

Table 7.1. Maximum emission reduction achievable for NO_x with measures included in the cost curves.. 105

Table 7.2. Maximum emission reduction achievable for NMVOC with measures included in the cost curves 106

Table 7.3. Scenarios selected for assessment... 109

Table 7.4. Resulting NO_x emissions from EU15 countries for each optimisation scenario 110

Table 7.5. Resulting NMVOC emissions from EU15 countries for each optimisation scenario 112

Table 7.6. Percentage of time for which ozone levels of 60-120 ppb are exceeded in the EU, as an average across all EU land-areas. 116

Table 8.1. Model results for the Greater Athens area for the reference case. 129

Table 8.2. Model results for the Greater Athens area – "business-as-usual" scenario 2010... 131

Table 8.3. Model results for the Greater Athens area for the scenario with 50 % reduction in local NO_x emissions on top of the "business as usual" scenario for 2010. 134

Table 8.4. Model results for the Greater Athens area for the scenario with 50 % reduction in local VOC emissions on top of the business as usual scenario for 2010. 134

Table 8.5. Model results for the Greater Stuttgart area for the reference case 138

Table 8.6. Model results for the Greater Stuttgart area for the business as usual scenario for 2010. 140

Table 8.7. Model results for the Greater Stuttgart area for the scenario with 50% reduction in local NOx emissions on top of the "business as usual" scenario for 2010. 140

Table 8.8. Model results for the Greater Stuttgart area for the scenario with 50% reduction in local VOC emissions on top of the "business as usual" scenario for 2010. 143

Table 8.9. Model results for the ozone levels in $\mu g/m^3$ at Milan metropolitan area for the reference case and the three future scenarios with meteorological conditions as in the period 2-3 July 1991. 147

Table 9.2. Source-receptor matrix 170

Table 9.3. Evolution of emission targets 171

Table 9.4. Cost savings 171

Table 9.5. Distributional impact (iteration 1) 172

Table 9.6. Remaining damages under policy scenario 185

Table 9.7. Reduction in damages under the policy scenario 187

Table 9.8. Gross Domestic Product 187

Table 9.9. Comparison of transfer payments under alternative principles 188

Source-Receptor-Matrix for AOT40 192

Source-Receptor-Matrix for AOT60 192

Table 11.1. Percentage of time for which ozone levels of 60-120 ppb are exceeded in the EU, as an average across all EU land-areas 207

Table 11.2. Comparison of modelled ozone concentrations in the base year 1990 and for the trend scenario 2010 207

Table 11.3. Emission reduction (relative to 1990) necessary to achieve given Gap Closures of AOTxx 209

Table 11.4. Total abatement costs for the selected scenarios for the EU15 209

Table 11.5. Number of days where 60 ppb is exceeded, and percentage of EU-area with more than 20 days of exceedance of 60 ppb per year 210

Table 11.6. Transfer payments under different burden sharing rules 216

1 Introduction

Rainer Friedrich

High ambient concentrations of ozone in the troposphere during sunny episodes are one of the main problems of air pollution in Europe. Ozone causes damage to human health, plants, animals and materials. Damage to human health includes irritation of respiratory tracts and eyes and reduction of lung function and physical power; in addition, ozone may cause or intensify chronic respiratory diseases. Furthermore, ozone decreases the photosynthesis rate in plants; this reduces for example crop yield and puts additional stress on forests. Materials containing carbons like paints, rubber and certain synthetics are affected.

The occurring damages are a strong signal that efforts should be made to reduce the ozone concentration in the lower troposphere. This is also reflected in currently existing or proposed ozone target values. As well the values given in the guidelines of the World Health Organisation (WHO) as the values proposed in the Ozone Directive, which has been adopted by the European Commission, are frequently exceeded in Europe. Thus, the need for a reduction of tropospheric ozone is obvious and a declared objective of European environmental policy.

However, it is a complex and difficult task to identify and to implement strategies for ozone reduction. The reason for problems with identifying such strategies lies in the complex and non-linear relationship between emissions of air pollutants and ozone concentrations. Ozone is not emitted, it is a secondary pollutant, that is formed in a complex chemical process in the atmosphere involving nitrous oxides, volatile organic compounds and sunlight. Parallel there are chemical processes that reduce ozone. While these chemical processes take place, the pollutants are transported over large distances and mixed vertically. So, ozone is not formed at the location of the emissions, but in a distance on the lee side. Apart from that, the spatial and temporal pattern of the release of precursors is different for the different pollutants. Furthermore there is a background concentration of ozone, so part of the ozone is transported from larger distances. With rising boundary layer in the morning , ozone from a reservoir layer is transported to the ground. All this underlines, that the formation of ozone is a highly complex non-linear process. So for identifying bundles of measures, that reduce precursor emissions in such a way, that the exceedance of ozone limit values is avoided, complex atmospheric transport and chemical transformation models have to be used. According to the transboundary nature of the ozone formation, the analysis of reduction strategies has to be made on the European level. On the other hand, the high spatial resolution of the processes involved requires a high spatial resolution of the atmospheric models.

The problem with implementing ozone reduction strategies is basically caused by the fact, that emission reduction measures may be quite expansive. So to achieve acceptance for abatement measures,

- it should be ensured, that the necessary emission reduction is carried out with the least costs and other impacts possible,
- the burden for the strategy, i. e. as well direct costs as macroeconomic impacts, should be shared in a fair way.

Minimising costs means that the costs and the potential for emission reduction of all relevant possible emission reduction measures have to be known and included into the optimising process. This process is especially difficult, as at the same time the location and the amount of emission reduction have to be varied to find the optimal solution.

In addition, costs are uncertain and may be higher or lower than estimated for the optimisation. So, the implementation of an ozone reduction strategy with an ecopolitical instrument, that is capable of improving the optimal solution taking into account the actual costs would be beneficial.

Burden sharing is especially important, as *burdens* (costs) and *benefits* (less damage due to reduced ozone concentrations) are not equally distributed among countries. This gives rise to reflections about possible burden sharing rules.

Regarding the necessity to reduce ozone concentrations on the one hand and the difficulties in identifying and implementing efficient strategies on the other hand, this book aims at providing recommendations and decision aid on how to implement an efficient strategy for the abatement of tropospheric ozone in Europe. The results are derived from applying a set of complex models, including emission, atmospheric and macroeconomic models. These models are also described here.

Chap. 2 of this book starts with a short general description of the formation of tropospheric ozone and the impacts that are caused by ozone. In addition, a number of ozone limit values is described. *Chap. 3* familiarises the reader with the main emission sources of the precursor pollutants, followed in *Chap. 4* by a description of the emission abatement measures, that are available or will become available within the time frame of the analysis, i. e. until 2010.

As all strategies to reduce ozone need time to be implemented, an analysis of possible strategies has to cover a time span, that reaches sufficiently far into the future to allow the market penetration of new technologies. So, as time horizon 2010 was chosen. This however makes it necessary to develop a base case scenario until 2010, that includes the development of the activities of the emission sources as well as the changes in emission factors caused e. g. by the implementation of emission abatement measures according to the current legislation (e.g. implementation of EURO 3 and EURO 4 thresholds for passenger cars). This scenario is described in *Chap. 5*. In *Chap. 6* the atmospheric model used for generating ozone concentrations in Europe is introduced and the ozone concentrations resulting from the emissions of the base case scenario are shown.

In addition, a sensitivity analysis reveals the relative importance of different emission source categories for the exceedance of ozone limit values.

Chap. 7 describes the optimising process of identifying efficient strategies to reduce ozone concentrations below certain limits. Results are sets of measures and the corresponding costs for each of the countries of the European Union.

However, as our model analyses the whole of Europe - with a grid size of ca. 50 km × 50 km- it is possible that small scale local effects are not seen. To investigate that, small scale (city) models are applied in addition that use the results of the European model as boundary condition and investigate the ozone concentration in three cities (Athens, Stuttgart and Milan) in detail (in *Chap. 8*).

Chap. 9 first describes an instrument, a new form of permit trading, that can be used to reduce the costs of achieving the desired reductions in tropospheric ozone concentration, if abatement costs are uncertain. Then, possible ways to achieve a fair burden sharing among the countries, that reduce their emissions, are explained. In *Chap. 10* a macroeconomic model is used to assess the macroeconomic impacts of the different scenarios. *Chap. 11* summarises the results and gives recommendations for implementing an efficient ozone reduction strategy in Europe.

2 Tropospheric Ozone

Stefan Reis and Rainer Friedrich

2.1
Tropospheric Chemistry and Ozone Precursor Substances

2.1.1
Ozone Formation in the Troposphere

Ozone is regarded to be the main photochemical oxidant present in the troposphere, originating from a formation process involving nitrogen oxides (NO_x), volatile organic compounds (VOC) and sunlight. VOCs are, according to the UNECE VOC-Protocol of 1991 (see *Sect. 2.3.2*) "... *all organic compounds of anthropogenic nature – other than methane – that are capable of producing photochemical oxidants by reactions with nitrogen oxides in the presence of sunlight* ...". They include pure hydrocarbons (containing only hydrogen and carbon) and organic compounds, containing further substances such as oxygen, nitrogen, chlorine or flourine.

This ozone formation process has been subject to thorough research for some time now, especially investigating the differing reactivities of the various hydrocarbons and their contribution to ozone production. Nitrogen oxides mainly originate from the combustion of fossil fuels, either in stationary sources (power and heat generation) or mobile sources (road transport). Besides anthropogenic emissions of hydrocarbons, e.g. from road transport or the use of organic solvents, biogenic emissions from trees and other plants are of importance.

In the presence of NO_2 and sunlight, *Eqns. 2.1* and *2.2* describe how ozone is formed. In the absense of VOCs, ozone is reduced again according to *Eqn. 2.3*.

$$NO_2 + h \cdot v \, (\lambda < 410 \, nm) \longrightarrow NO + O(^3P) \quad \text{(E 2.1)}$$

$$O_2 + O(^3P) \longrightarrow O_3 \quad \text{(E 2.2)}$$

$$NO + O_3 \longrightarrow NO_2 + O_2 \quad \text{(E 2.3)}$$

Thus, NO, NO$_2$ and ozone concentrations form a stable equilibrium, which would not lead to such high ozone concentrations as they are currently observed. However, if VOCs are present, the following reactions (*Eqns. 2.4 and 2.5*) occur.

$$OH + RH \ (+O_2) \longrightarrow RO_2 + H_2O \qquad (E\ 2.4)$$

$$RO_2 + NO \longrightarrow NO_2 + RO \qquad (E\ 2.5)$$

RH is the representation of VOCs, in which H represents a hydrogen and R the rest of the molecule, e.g. CH$_3$, C$_2$H$_5$ or other alkyl radicals. As NO$_x$ is oxidised according to *Eqn. 2.4*, it cannot reduce ozone (*Eqn. 2.3*), leading to an increase in ozone concentrations. The oxidation of NO to NO$_2$ by RO$_2$ or HO$_2$ molecules and subsequently the formation of O$_3$ from NO$_2$ and sunlight is described in more detail in *Fig. 2.1*.

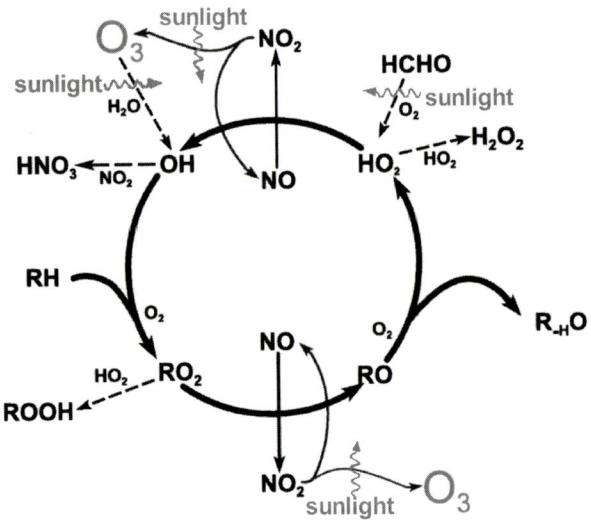

Fig. 2.1. Tropospheric cycles of catalysis (Source: PORG, 1997)

NO$_x$ controlled and VOC controlled regions can be distinguished, where ozone formation is limited by the respective lack or surplus of one of the components. In urban areas, where a large amount of NO$_x$ is emitted, ozone levels are most often low because of the reaction described in *Eqn. 2.3*. In rural areas, some distance downwind of NO$_x$ sources, the direct NO-sink becomes less important and a more balanced ratio between ozone precursors often leads to ozone generation and peak concentrations. For a more extensive discussion of ozone chemistry, including the important sink processes of HNO_3 and H_2O_2 formation, see, for example, Seinfeld and Pandis (1998).

VOCs are no homogenous substance group, but a collection of a vast number of volatile organic compound species with different chemical properties and reactivities. A first distinction is usually made between methane, which has a low reactivity with OH (*Eqn. 2.4*) and is thus less important for short term ozone formation, and the so-called NMVOCs. However, NMVOC still consists of a large number of species with highly different physical and chemical characteristics and different ozone forming potentials. Given such differences, which can be accounted for to some extent with atmospheric chemical models, it is clear that an assessment of the NMVOC speciation associated with each source is important to assessing the effects of control measures. For example, hydrocarbons, like toluene, ethene, butane and propene are mainly responsible for short-term ozone creation, while slower reacting alkanes play a more important role, when long-term (e.g. multi-day) formation is regarded (Andersson-Sköld et al. 1992).

Though, current emission inventories usually do not account for different species of NMVOC. For this study, the speciation was done in the stage of calculating spatial high resolution emission data for the disperson models, using the speciation developed by Middleton and Carter (1999) and work conducted at IER (Friedrich and Obermeier 1999).

2.1.2
Anthropogenic Emissions of Ozone Precursor Substances

In order to develop strategies for ozone abatement, it is vital to know the structure and shares of all emission sources of ozone precursor substances, namely NO_x and NMVOC. In this case, the CORINAIR[1] emission inventory for Europe was taken as a basis.

This inventory provides a detailed collection of emission data, distinguishing the main sectors of activity relevant to air pollutant emissions. These SNAP[2] groups are usually subdivided down to activity level and present the best currently available information base for European emission data. *Figs. 2.2* to *2.5* show the contribution of different source sectors (for the 15 European Union Member States) to total emissions of NO_x and NMVOCs for CORINAIR 94, the most recent inventory, and maps of the spatial distribution of emissions over Europe (for high resolution maps cf. *Plate 9*). The detailed analysis of the emission sources is described in *Chap. 3*. For NO_x, three relevant source sectors can be identified, *stationary combustion, road transport* and *other mobile sources*.

---

[1] CORINAIR: Emission inventory for air pollutants developed under the work programme CORINE (***CO**-o**R**dination d'**IN**formation Environmentale*) according to Council Decision 85/338/EEC (OJ, 1985); the first CORINAIR emission inventory was compiled for the year 1985 and covered three air pollutants (SO_2, NO_x and VOC)

[2] SNAP: **S**elected **N**omenclature for **A**ir **P**ollution, nomenclature developed in the frame of CORINE to relate emissions of air pollutants to relevant source sectors, sub-sectors and activities

In the case of NMVOCs, the picture is somewhat different with *road transport*, *solvent use* and biogenic emissions being the main source categories.

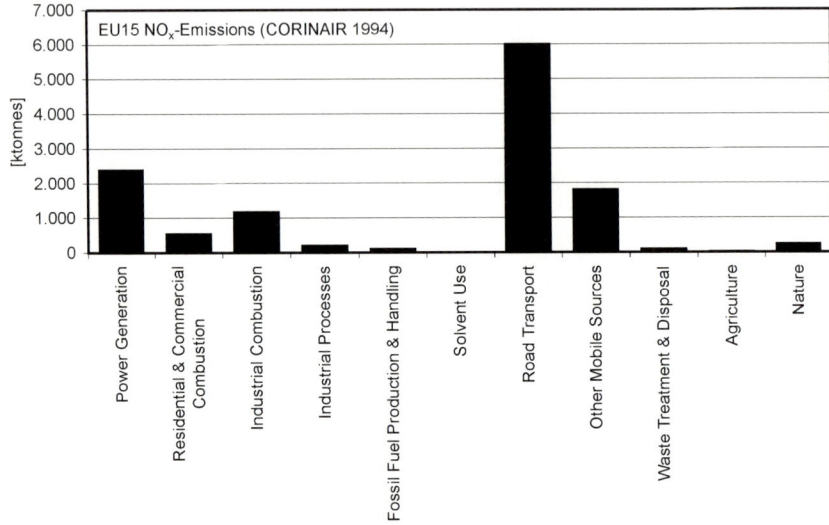

Fig. 2.2. Sectoral contributions to emissions of NO_x for the EU_{15} in 1994 (Source: CORINAIR 94)

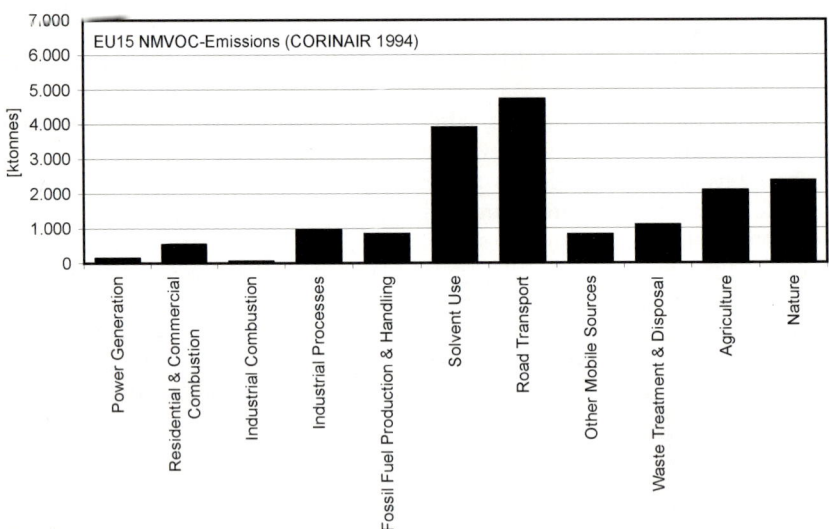

Fig. 2.3. Sectoral contributions to emissions of NMVOC for the EU_{15} in 1994 (Source: CORINAIR 94)

2.1 Tropospheric Chemistry and Ozone Precursor Substances

The spatial structure of both NO_x and NMVOCs is shown in *Figs. 2.4* and *2.5*.

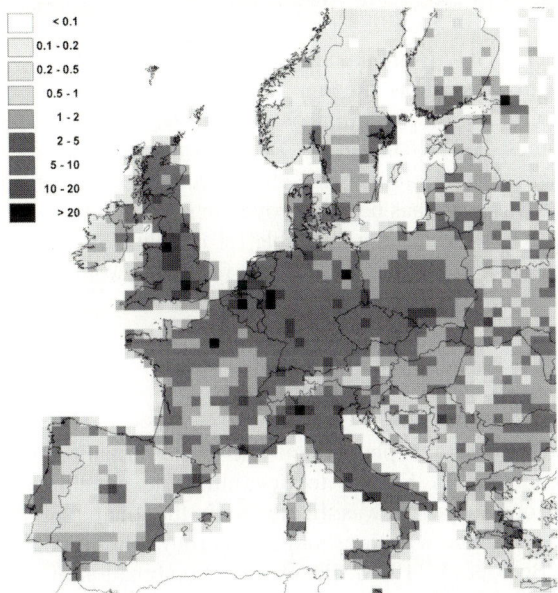

Fig. 2.4. Spatial distribution of NO_x emissions in Europe 1994 in (tonnes/km^2) (Source: Ebel et al. 1997)

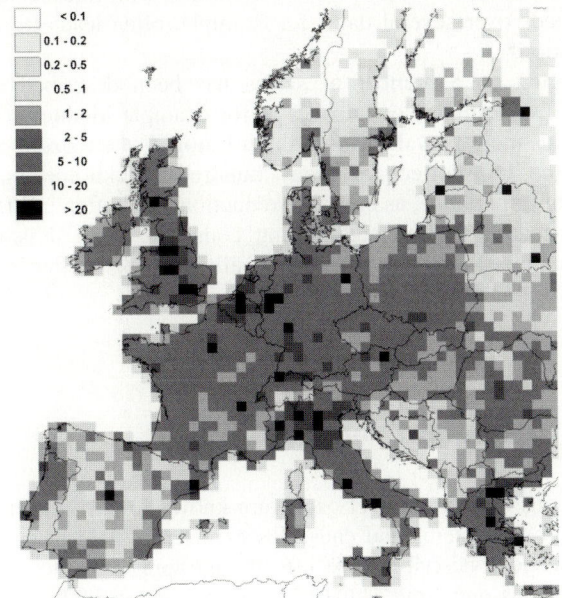

Fig. 2.5. Spatial distribution of NMVOC emissions in Europe 1994 in [tonnes/km^2] (Source: Ebel et al. 1997)

2.1.3
Biogenic Emissions of Ozone Precursors

While emissions of NO_x from natural sources are probably low (though very uncertain), biogenic emissions amount to approximately 30 % of total EU15 NMVOC emissions according to Simpson et al. 1999. Though, emission estimates for biogenic sources are still subject to vast uncertainties, since the data situation on emission potentials and biomass are still scarce, even though the quality and availability of land-use data has been improved in recent years.

Finally, these emissions are not available for applying abatement measures, so remaining NO_x emissions could lead to ozone formation even if no anthropogenic NMVOC would be emitted.

2.1.4
Meteorology and Atmospheric Dispersion

The meteorological conditions play a vital role in the formation of ozone. Several studies were conducted in Europe to determine the conditions in which the occurrence of ozone episodes was likely, with the final aim to improve the predictability of exceedances of thresholds. A thorough discussion of these studies can be found in Beck et al. (1998). Basically, a combination of surface pressure, surface temperature and solar radiation, along with the horizontal wind field have been identified to be the main parameters to determine the development of ozone episodes. A stable summerly high pressure condition with intense solar radiation at low wind speeds over several days, for example, often leads to a build up of ozone concentrations

Ozone formation on a number of scales has been described in numerous studies, ranging from single-day episodes in for example Mediterranean cities to multi-day episodes more prevalent in northern Europe, where ozone and/or ozone precursors can be transported for many hundreds of kilometers. This latter phenomenon is of importance, as emission reductions will often influence ground-level ozone concentrations in neighbouring countries. Thus it is necessary to develop a harmonised strategy including all emitting countries to reduce exceedances of ozone thresholds all over Europe.

2.2
Impacts on Health and Vegetation

High concentrations of ground level ozone are known to have adverse effects on human health and ecosystems and cause losses in crop yield of sensitive crops. The following sections describe these effects and approaches to assess impacts and derive thresholds and limit values for the protection of human health and crops and other vegetation.

2.2.1
Effects on Human Health

The occurrence of adverse effects of increased concentrations of ground level ozone on human health have been widely accepted, even though it is difficult to define an overall level of impact for the whole population, as ozone peak concentrations usually occur localized. It has been noted, that people with asthmatic or similar respiratory diseases are especially sensitive to ozone concentration increases. *Table 2.1* shows an overview of expected acute health impacts of different ambient concentrations.

Table 2.1. Health impacts of different ambient concentrations of ground level ozone

Ozone Concentration ($\mu g/m^3$)	Exposition	Health Impact
40	30 minutes	smell noticeable
100	30 minutes	beginning irritation of the respiratory tract
160	6.6 hours	beginning impairment of lung function
240	30 minutes	reduced physical fitness
620	15 minutes	considerable irritation of the respiratory tract
1000	1 hour	stinging eyes, claustrophobia
2000	30 minutes	severe impairment of lung function

Source: BMU (1995)

2.2.2
Impacts on Agricultural Crops and Forests

Besides the health effects described above, high concentrations of ground level ozone can seriously damage sensitive plants like agricultural crops or forest trees.

For agricultural plants, the damage can be observed as reduced crop yields, leading to an immediate monetary loss. After long-term investigations of impacts of air pollution on agricultural crops had been conducted first in the *US (National Crop Loss Assessment Network*, Somerville et al. 1989), a European research programme was set up called the *European Open-Top Chamber (OTC) Programme* (cf. Skärby et al 1993).

As a result of this programme, yield curves were calculated depending on the ambient concentration of ground-level ozone, showing, that current ozone levels in Europe have a significant impact on crop yield of wheat, while barley and oats were less sensitive to ozone exposure. Ozone effects on forest trees vary with different sensitivities of tree types. Recently this type of research has made use of the AOT40 concept (see *Sect. 2.2.3*), and established very good exposure-response relationships for OTC experiments (*Fig. 2.6*).

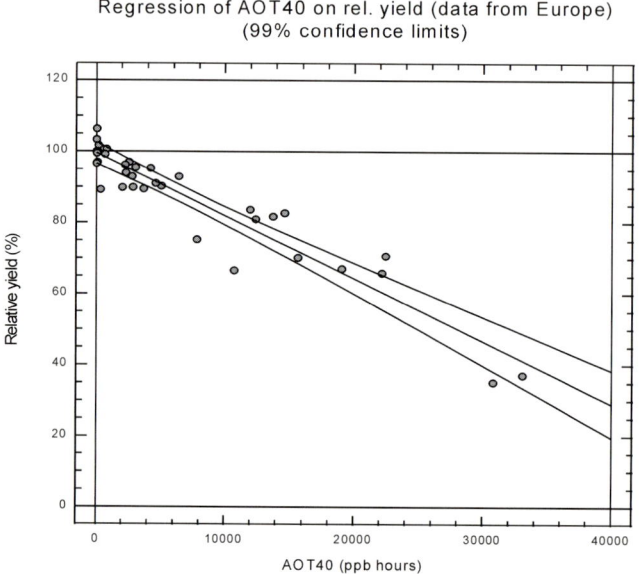

Fig. 2.6. Yield-loss versus AOT40 relationship (Source: Pleijl 1996)

2.2.3
The Concept of Critical Levels and Loads

The concept of critical levels and loads was developed in the frame of the UNECE[3] *Convention on Long Range Transboundary Air Pollution* (CLRTAP, *see Sect. 2.3.2*) which dates back to the year 1979. They have been defined as follows (UNECE 1996):

- **Critical Levels** are atmospheric concentrations of air pollutants, *above* which adverse effects on receptors such as plants, ecosystems or materials can be expected according to the state of the art.
- **Critical Loads** are quantitative assessments of the exposition to one or more pollutants, *below* which – according to the state of the art – no significant adverse effects to specified sensitive parts of the environment can be expected

Defining the exact limits for these critical levels/loads has been subject to thorough and controversial scientific discussion. In the case of tropospheric ozone an additional question occurred, if either the seasonal mean or the summation of

---

[3] United Nations Economic Commission for Europe

exceedances (e.g. AOTxx) of a specified threshold should be selected to assess the impacts on plants and ecosystems. The latter will be extensively used in this book and is described in *Sect. 2.2.4*. *Table 2.3* shows the critical levels set by the UNECE.

2.2.4
Source-Receptor Relationships

Along with the concept of critical levels/loads, the approach of using source-receptor relationships will be used later. Source-receptor matrices give ratios of how the concentration of a pollutant changes for each receptor (e.g. country, grid cell) for each unit the emissions of its precursors in- or decrease at the source (country, grid cell).

Having been used for modelling acidification first (Alcamo et al. 1990), the approach had to be adapted to be used for tropospheric ozone, because the relation between the emissions of NO_x and NMVOC and the concentration of ozone is non-linear. In NMVOC controlled areas, though, ozone formation seems to be almost in linear dependence on changes in NMVOC concentration changes.

2.2.5
Limit Values and Cumulative Approaches

Limit values for ozone concentrations are generally given in $\mu g/m^3$ (1 μg = 10^{-6} g) or in ppb[4] (parts per billion), typically referring to averages over 1 hour to 24 hours, up to annual means. These limit values are used to assess peak concentrations and provide thresholds for public information, warning or taking action (e.g. traffic bans during ozone episodes). To assess the long-term development of ozone concentrations, cumulative approaches are applied to calculate the number of exceedances of a threshold. These thresholds are defined with respect to the objective, either health protection, or the protection of crops or ecoysystems.

For the protection of vegetation, European critical levels are usually expressed as an **A**ccumulated exposure **O**ver a **T**hreshold of **40** ppb, in short AOT40. For daylight hours with global radiation $\geq$ 50 Wm^{-2} the sum of differences between hourly concentrations of ozone above 40 ppb is calculated according to the equation:

$$AOT40 = \sum_{i=1}^{n} [c_{O_3} - 40]_i \text{ for } c_{O_3} > 40\, ppb \quad [unit: ppb.h]$$

c_{O_3}: hourly ozone concentration in ppb
i : running index
n : number of hours with $c_{O3} > 40$ ppb

---

[4] at 20 °C and 1013 mb pressure, 1 ppb equals 2.00 $\mu g/m^3$

AOT40 is calculated for agricultural crops, natural and semi-natural ecosystems (over three months, May to July) and forest trees (over six months, April to September), covering the periods where ozone concentrations are likely to have adverse effects on plant growth and crop yield.

Regarding health protection, the main concern is the peak concentrations, which are difficult to assess by using a cumulative approach. Though, AOT60 is widely accepted as a proxy value. AOT60 is calculated in analogy to AOT40 over a period of 6 months (April to September). A level of 0 ppb.h of AOT60 is mathematically equivalent to zero exceedances of the WHO guideline of 120 µg/m^3.

2.3
Air Quality Legislation, Thresholds and Standards

2.3.1
Legislative Overview

A vital aspect of air pollution is the scope of its impact. Some air pollutants have mainly local effects, others are dispersed by atmospheric transport and thus have regional and transboundary effects. Within the last 20 years, this transboundary scope of air pollution as been more and more acknowledged and supra-national bodies like the UNECE, the EC and finally the United Nations Framework Convention on Climate Change (UNFCCC) have taken the lead in the development of harmonised approaches to improve air quality.

Since the ozone itself and its precursors NO_x and especially NMVOCs have relatively long lifetimes in the atmosphere, ozone concentrations in each European Member State depend not only on the emissions of precursors within the country itself, but on *imported* air pollution from other countries. Thus, reduction of ground-level ozone is only possible if joint action is taken in all countries contributing to the problem.

The UNECE has undertaken an approach of negotiating agreements under *the Convention on Long Range Transboundary Air Pollution (CLRTAP),* leading to protocols for all relevant air pollutants (see *Sect. 2.3.2*).

Current developments, such as the EC Air Quality Framework Directive or the multi-effect, multi-pollutant protocol on nitrogen oxides and related substances being prepared by the UNECE, addressing photochemical pollution, acidification and eutrophication, reflect the necessity for an integrated assessment of air pollution abatement. The interdependency of both emission sources and environmental effects of emissions has led to the conclusion that in order to find an optimal and efficient pathway to improve air quality, concepts for the reduction of all relevant air pollutants have to be developed in a harmonised way.

Finally, the 5th Environmental Action Plan (5EAP) of the EU sets targets for the reduction of emissions of NO_x (30 % by the year 2000 from 1990 emission levels and NMVOCs (stabilisation by 1994 and 30 % reduction by the year 2000, both from 1990 emission levels). In addition to that, the EU Ozone Abatement Strategy

2.3 Air Quality Legislation, Tresholds and Standards

is currently being improved by the development under the Daughter Directive on Ozone under the EU Framework Directive.

Table 2.2. Protocols to the CLRTAP (Source: UNECE 1996)

Protocol	Year	Emission Target	Target Year
1st Sulfur Protocol	1985	30% reduction of total sulfur emissions or their transboundary fluxes on the basis of 1980 emissions	1993
NO$_x$ Protocol	1988	1st step: stabilisation of 1987 (exception: USA 1978) emissions of NOx 2nd step: application of an effects-based approach to further reduce NO$_x$ emissions or their transboundary fluxes	1987
VOC Protocol	1991	30% reduction in emissions of volatile organic compounds (VOCs) by 1999 using a year between 1984 and 1990 as a basis	1999
2nd Sulfur Protocol	1994	An effects-based approach, the critical load concept, best available technology, energy savings, the application of economic instruments and other considerations was applied in the preparation of the Protocol. This has led to a differentiation of emission reduction obligations of Parties to the Protocol. The effects-based approach, which aims at gradually attaining critical loads, sets long-term targets for reductions in sulphur emissions, although it has been recognised that critical loads will not be reached in one single step	–
Heavy Metals Protocol	1998	reduce emissions for cadmium, lead and mercury below their levels in 1990 (or an alternative year between 1985 and 1995), targeting emissions from industrial sources (iron and steel industry, non-ferrous metal industry), combustion processes (power generation, road transport) and waste incineration	–
Protocol on Persistent Organic Pollutants (POPs)	1998	objective is to eliminate any discharges, emissions and losses of POPs, among others to reduce emissions of dioxins, furans, PAHs and HCB below their levels in 1990 (or an alternative year between 1985 and 1995) and laying down limit values for the incineration of municipal, hazardous and medical waste	–
Multi-effect-multi-pollutant Protocol	*in prep.*	addressing ground-level ozone, acidification and eutrophication as the effects of emissions of several air pollutants, e.g. NO$_x$, NMVOCs, SO$_2$, NH$_3$ etc.	–

2.3.2
United Nations Economic Commission for Europe (UNECE)

With the 1979 *Convention on Long Range Transboundary Air Pollution*, the UNECE implemented the first multilateral treaty with the target of reducing air pollutant emissions and thus improving European air quality. One of the key activities within the CLRTAP is *the Co-operative Programme for Monitoring and Evaluation of the Long Range Transmission of Air Pollutants in Europe* (EMEP). EMEPs prime objective is to provide data on depositions, concentrations, long-range transport and transboundary fluxes of air pollutants.

With several protocols to the CLRTAP being in force and ratified by a large number of European countries (see *Table 2.2*), a multi-effect protocol is now under preparation, which is designed to take an integrated approach to a number of pollutants and effects. Within the different protocols, limit values for air pollutants have been set as well as reduction targets and deadlines to achieve these targets.

Table 2.3. Critical Levels for Vegetation

Pollutant	Unit	Forests	Ecosystems
SO_2	$\mu g/m^3$	**20** *annual mean* and *winter mean* October to March	
O_3	ppm.h	**10** *AOT 40^a over 24h and 6 months*	**5.3** *AOT 40^a over day hours and 3 months*
NO_x	$\mu g/m^3$	**30** *annual mean* **95** *4 h mean*	
NH_3	$\mu g/m^3$	**3300** *1 h mean* **270** *24 h mean* **23** *1 monthly mean* **8** *annual mean*	

^{a} Accumulated exposure **O**ver **T**hreshold 40 ppb *(see 2.2.4)*   (Source: UNECE 1996)

2.3.3
World Health Organisation (WHO)

The WHO focuses on health impacts of various substances, developing limit values and critical levels. In the second edition of the WHO air quality guidelines in 1996, an ozone concentration of 120 $\mu g/m^3$ as a maximum of an 8 hour average was set as a guideline value for health protection. For NO_x, 200 $\mu g/m^3$ as a 1 hour mean, respectively 40 $\mu g/m^3$ as an annual mean were set, while only few limit values for specific hydrocarbons were included.

The threshold for the protection of agricultural crops (5 % yield loss), natural and semi-natural vegetation is set to an AOT40 of 3 ppm.h, for the protection of forest trees to an AOT40 of 10 ppm.h.

2.3.4 European Union

The European Union has recently introduced a new approach to tackling air quality issues, moving away from addressing each pollutant or problem independently towards an integrated air quality management. With the development of the Air Quality Framework Directive (Directive 96/62/EC) on ambient air quality assessment and management basic principles have been set to deal with air pollution on a European scale. The aims of this directive can be summarised as follows:

- defining and establishing ambient air quality objectives to avoid, prevent or reduce harmful effects on human health and the environment as a whole
- assessing ambient air quality in Member States on the basis of common methods and criteria
- producing adeaquate publicly available information about ambient air quality and
- maintaining ambient air quality where it complies with targets and improve, where it does not comply.

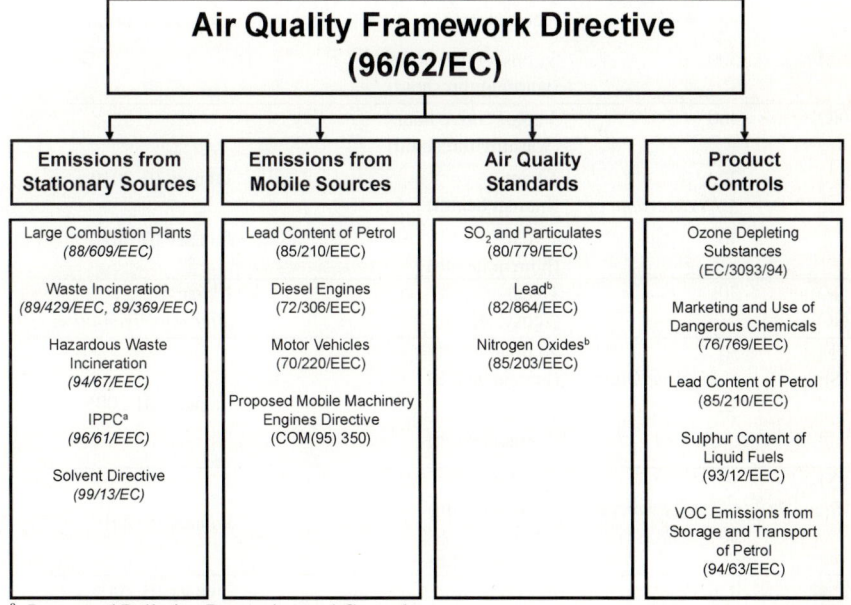

[a] Integrated Pollution Prevention and Control
[b] to be replaced by new standards under the Air Quality Framework Directive, which also introduce standards for ozone (see *Table 2.7*), CO, benzene and other atmospheric pollutants (see *Table 2.5*)

Fig. 2.7. Areas covered by the European Community's Air Quality Framework Directive

The European Community has set out to establish optimal ambient air quality limit values, as well as assessment procedures and reporting requirements within ten to fifteen years through daughter directives covering specific individual pollutants or groups of pollutants. Previous directives, e.g. concerning sulphur dioxide, particulates or nitrogen oxides are being replaced by these daughter directives.

For tropospheric ozone, the EC has issued the Council Directive on Air Pollution by Ozone in 1992 (92/72/EEC), which sets threshold concentrations (see *Table 2.4*) and establishes monitoring and data exchange procedures for informing and, if necessary, warning the population.

Table 2.4. Limit Values of the Air Quality Framework Directive (as to May 1999)

pollutant	limit value ($\mu g/m^3$)	reference period (targeted at the protection of ...)	target date to comply with limit value
SO_2	350 ≤ 24 *exceedances*	1 hour (human health)	January 01 2005
	125 ≤ 3 *exceedances*	24 hours (human health)	January 01 2005
	20	year/winter period (ecosystems)	1 year *after implementation*
	500	3 consecutive hours (warning threshold)	–
NO_2	400	3 consecutive hours (warning threshold)	–
	200 ≤ 18 *exceedances*	1 hour (human health)	January 01 2010
	40	1 year (human health)	January 01 2010
$NO + NO_2$	30	1 year (ecosystems)	1 year *after implementation*
PM_{10} (Stage 1)	50 ≤ 35 *exceedances*	24 hour (human health)	January 01 2005
	40	1 year (human health)	
PM_{10} (Stage 2)	50 ≤ 7 *exceedances*	24 hours (human health)	January 01 2010
	20	1 year (human health)	
lead	0,5	1 year (human health)	January 01 2005
benzene	5	1 year (human health)	January 01 2010
CO	10 mg/m^3	8 hour mean (human health)	January 01 2005

2.3 Air Quality Legislation, Tresholds and Standards

Member States have to report the results of continuous measurements of ozone concentrations to the European Commission, providing the following information:

- maximum, median and 98th percentile values of 1 hour and 8 hour mean concentrations
- occurrence of exceedances of threshold levels (number, date and duration of the episode)
- peak concentration monitored in an episode of exceedance (Beck et al. 1996)

The population information threshold of 180 μg/m^3 was exceeded several times per year in most of the EC Member States, while the population warning threshold (360 μg/m^3, 1 hour mean) was attained in a few situations. The information that was issued when a threshold was exceeded received considerable attention from the media, and the increased public awareness has brought the ozone problem higher on the list of political priorities.

Table 2.5. Ozone threshold values set by the EC Directive on Air Pollution by Ozone

Description	Criteria based on	Value
population information threshold	1 hour average	180 μg/m^3
population warning threshold	1 hour average	360 μg/m^3
health protection threshold	fixed 8 hour means; period hours: 0:00-8:00, 8:00-16:00, 16:00-24:00, 12:00-20:00	110 μg/m^3
vegetation protection threshold	1 hour average	200 μg/m^3
	24 hour average	65 μg/m^3

Within the Air Quality Framework Directive, a new directive concerning ground-level ozone is in preparation, the so called *Daughter Directive* on Ozone, which shall replace Directive 92/72/EEC. This pending Daughter Directive gives a combination of long-term objectives, target values and alert thresholds, as outlined in *Table 2.7*. The Daughter Directive is currently being assessed on the basis of a proposal by the EC from early 1999 and will probably be adopted within the next year.

Table 2.6. Ozone target values and objectives set by the Draft Daughter Directive on Ozone

Description	Averaging Period	Value	Target Year
long term objective, human health	yearly maximum 8 h mean (rolling average)	120 µg/m³	–
long term objective, vegetation	AOT40, May to July between 8.00 and 20.00 CET	6 000 µg/m³	–
target value, human health	daily maximum 8 h mean (rolling average)	120 µg/m³ 20 – 25 exceedances	2010
information threshold	1 hour	180 µg/m³	–
alert threshold	1 hour	240 µg/m³	–

2.4 Tropospheric Ozone – a Local or Regional Problem?

A final problem to solve is to determine the optimal scope of a study like this. It is necessary to cover the whole region, in this case Europe, in order to account for transboundary transport of pollutants and effects, which are especially important for generating the high levels of ozone found in rural areas in northern Europe. On the other hand, especially in some Mediterranean cities, ozone episodes sometimes affect a comparatively small area, with peak concentrations occurring downwind of an urban area or at individual sites only. These peaks cannot be properly resolved by regional models, which use a grid resolution of 50 to 150 km. For this study, a hybrid approach was chosen, using the EMEP model (*see Chap. 6*) at a resolution of either 50 or 150 km to cover the regional (macroscale) aspects and two mesoscale models to investigate three selected urban areas (*see Chap. 8*). The mesoscale model used boundary conditions calculated by the macroscale model. This way, the overall situation of tropospheric ozone concentrations over Europe could be assessed as well as the specific problems of ozone peaks in the vicinity of conurbations.

2.4.1 Local Ozone Formation and Transboundary Transport

The aspect of transboundary dispersion of ozone precursors has already been addressed in some of the previous sections. Though, it is one of the crucial points when trying to develop harmonised abatement strategies, since cause and effect – emissions and ozone formation– do not always occur within the same country. The UNECE CLRTAP has acknowledged this, using negotiating processes to achieve agreements which can be accepted by all countries affected, those who have to carry the burdens of reducing emissions and those who have the benefits from reduced concentrations of air pollutants and their effects. The approach towards burden sharing taken in this study is described in *Chap. 9*, while the most important information about the source-receptor relations are provided by the EMEP model. Using these tools, it was possible to determine the contribution of each countries' emissions to ozone formation in all other countries (in fact, EMEP grid cells) and thus derive mechanisms to account for inequalities between countries that have to spend comparatively more than they would gain. Within these matrices, the contribution to ozone formation within the country itself was given as well. For some countries (e.g. The Netherlands and the United Kingdom), the effect of their NO_x emissions on neighbouring countries is even negative, meaning a decrease in emissions of NO_x could lead to an increase of ozone concentrations.

2.4.2
Urban and Rural Areas

As described in *Sect. 2.1*, ozone formation depends on a number of conditions. One of these prerequisites is the ratio between NO_x and hydrocarbon concentrations, limiting the amount and, depending on the type of hydrocarbons, the rate of ozone creation.

A characteristic feature of conurbations is a high traffic density and localised power plants and industrial areas. Furthermore, access roads and motorways in and around these urban areas with comparatively high shares of freight traffic contribute to an overall excess of NO_x emissions in relation to NMVOC. The high NO emissions density suppresses ozone levels inside cities (*eqn. 2.5*), thus, ozone peaks usually occur downwind of conurbations, where the high NO_x and NMVOC concentrations of the urban plume mix with and are diluted by air from the surrounding rural area (often rich with hydrocarbons, e.g. biogenic emissions from forests) and create favourable conditions for ozone formation.

This aspect is in so far important as reductions in NO_x emissions only might well lead to an increase in ozone concentrations, especially in the suburbs of larger conurbations, exposing even more people to increased levels of tropospheric ozone. Therefore, reduction strategies have to be designed with great care, ensuring a simultaneous and balanced decrease of anthropogenic emissions of either NO_x and NMVOC, while taking into account the contribution of biogenic sources of hydrocarbons.

2.4.3
Exceedances of Thresholds and Limit Values

In order to monitor concentrations of ground-level ozone, the EC Directive on Air Pollution by Ozone included requirements for a European measuring network, with measuring stations in specified spatial distribution, covering urban as well as rural locations. As indicated in *Fig. 2.6*, the number and distribution of these stations is not yet capable of monitoring all parts of the European Community. Germany, Austria, Belgium, the Netherlands and Luxembourg are fairly well covered, while in the southern European countries monitoring sites are more or less scarce.

According to the EEA, the threshold for human health (110 µg/m^3 as an 8 hour mean) was exceeded substantially in the years 1995 and 1996 (de Leuw and van Zantvoort 1996) with an estimated 41 million EU15 inhabitants experiencing exceedances, 80 % of these of more then 25 days. Beck et. al. (1998) state that it is still difficult to assess the short-term impacts of increased ozone concentrations due to the lack of data, but increased hospital admissions and other direct effects on people with respiratory problems are widely acknowledged. To quantify the direct effects of exceedances of ozone thresholds on human health, though, more studies on these effects are needed.

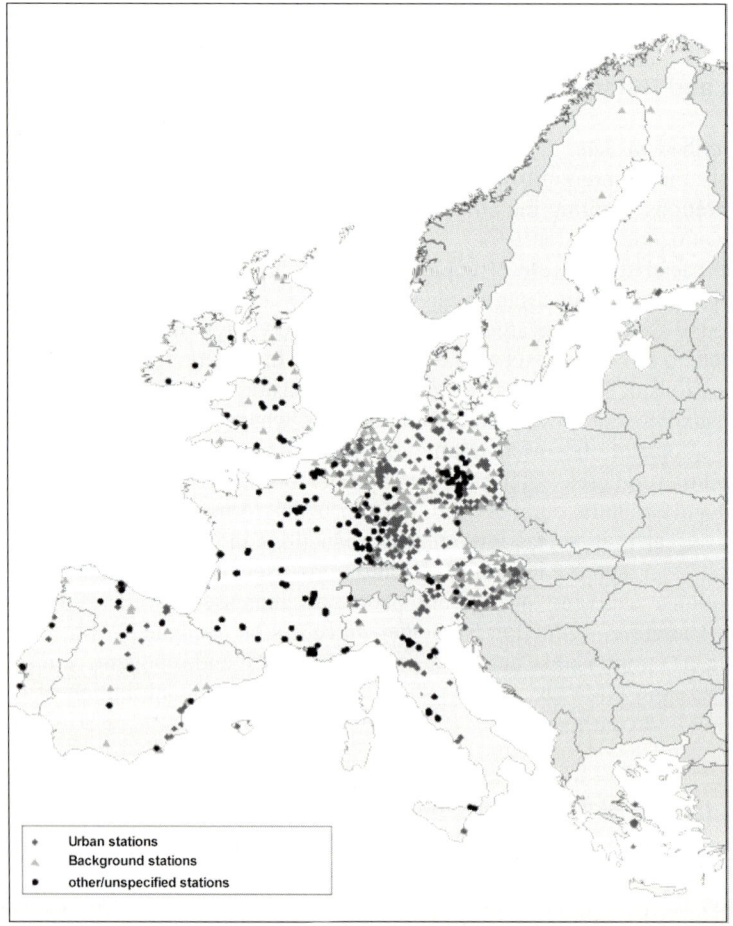

Fig. 2.8. European network of measuring stations for tropospheric ozone (Source: ETCAQ)

2.5 References

Alcamo J, Shaw R, Hordijk L (1990) The RAINS Model of Acidification. International Institute for Applied Systems Analysis IIASA, Kluwer Academic Publishers, Dordrecht

Andersson-Sköld Y, Grennfelt P, Pleijel K (1992) Photocemical Ozone Creation Potentials – A Study of Different Concepts. in: Journal of the Air and Waste Management Association, 42, pp 1152–1158

Beck J P, Krzyzanowski M, Koffi B (1998) Tropospheric Ozone in the European Union – The Consolidated Report. EEA, Copenhagen

BMU http://www.eea.dk (1995) UMWELT: Informationen des Bundesumweltministeriums. Issue 9/1995, Bundesumweltministerium, Berlin

Derwent R G, Jenkin, M E, Saunders S M (1996) Photochemical Ozone Creation Potentials for a Large Number of Reactive Hydrocarbons Under European Conditions. in: Atmospheric Environment, 30 (2) pp. 181–199, Elsevier Science

ETCAQ: http://www.etcaq.rivm.nl/

EUROSTAT (1995) Europe's Environment: Statistical Compendium for the Dobříš Asseessment. ECSC-EC-EAEC, Brussels Luxembourg

Friedrich R, Krewitt W (1997) Umwelt- und Gesundheitsschäden durch dir Stromerzeugung – Externe Kosten von Stromerzeugungssystemen. Springer, Berlin Heidelberg New York Tokio

Friedrich R, Obermeier A (1999) Anthropogenic Emissions of Volatile Organic Compounds. in: Hewitt C N ed (1999) Reactive Hydrocarbons in the Atmosphere, Acadamic Press, San Diego, pp 2 – 38

Ebel A, Friedrich R, Rodhe H eds (1997) Tansport and Chemical Transformation of Pollutants in the Troposphere – Volume 7: Tropospheric Modelling and Emission Estimation. Springer, Berlin Heidelberg New York Tokio

Hov Ø ed (1997) Tropospheric Ozone Research – Tropospheric Ozone in the Regional and Subregional Context. Springer, Berlin Heidelberg New York Tokio

Hewitt C N ed (1999) Reactive Hydrocarbons in the Atmosphere. Acadamic Press, San Diego

Lefohn A S (1994) Surface ozone. Pergamon Press, Oxford

de Leuw F, van Zantvoort E (1996) Exceedance of Ozone Threshold Values in the European Community in 1995. EEA, Copenhagen

Middleton P, Carter W P L, Stockwell W R (1990) Aggregation and Analysis of Volatile Organic Compound Emissions for Regional Modelling. in: Atmospheric Environment, Vol 24 A No 5, pp 1107–1133

Pleijl, H., Statistical aspects of critical levels for ozone based on yield reductions in crops (1996), In Kären-lampi and Skärby (eds.), Critical Levels for Ozone in Europe: Testing and Finalizing the Concepts, (1996), UN-ECE Workshop Report, Univ. of Kuopio, Dept. of Ecol. and Environ.Sci., Finland.

PORG (1997) Ozone in the United Kingdom – Fourth Report of the Photochemical Oxidants Review Group (PORG). UK Dept. Environment, Transport and the Regions, London

Seinfeld J H, Pandis S N (1998) Atmospheric Chemistry and Physics. John Wiley & Sons, New York

Simpson, D., Winiwarter, W., Börjesson, G., Cinderby, S., Ferreiro, A., Guenther, A.,Hewitt, C. N. , Janson, R., Khalil, M. A. K., Owen, S., Pierce, T. E., Puxbaum, H., Shearer, M., Skiba, U., Steinbrecher, R., Tarrason, L., and Öquist, M. G., 1999, Inventorying emissions from nature in Europe, J. Geophys. Res., 104, No. D7, 8113-8152.

Skärby L, Selldén G, Mortensen L et al (1993) Responses of Cereals Exposed to Air Pollutants in Open-Top Chambers. in: Jäger H J, Unsweorth M H, De Temmermann L, Mathy P (eds) Effects of Air Pollution on Agricultural Crops in Europe, Air Pollution Research Report 46, European Commission, Brussels, pp 241–259

Somerville M C et al (1989) Impact of Ozone an Sulfur Dioxide on the Yield of Agricultural Crops. Technical Bulletin 292, North Carolina Agricultural Research Service, Raleigh

Stanners D, Bourdeau P (1995) Europe's Environment – The Dobříš Asseessment. European Environment Agency, Copenhagen

UNECE (1996) Manual on Methodologies and Criteria for Mapping Critical Levels/Loads and Geographical Areas Where They are Exceeded. UBA Texte 71/96, Umweltbundesamt, Berlin

UPI (1999) Neue medizinische Erkenntnisse über die gesundheitlichen Auswirkungen von Sommersmog, Berechnung der durch Ozon verursachten Todesfälle in der Bundesrepublik Deutschland. UPI Report No. 47, Umwelt- und Prognose-Institut e.V., Heidelberg

3 Emissions of Ozone Precursors

Stefan Reis and Rainer Friedrich

3.1
Source Sector Analysis

A brief introduction into the sources of ozone precursor emissions has already been given in *Sect. 2.1.2*. A detailed analysis of these sectors will be conducted in the following sections, providing the basis for the selection of suitable abatement measures and being used for the projection of a future emission development.

3.1.1
Sources of NO_x Emissions

NO_x emissions originate almost exclusively from combustion of fossil fuels. As well nitrogen from the air used for combustion (thermal and prompt NO) as nitrogen contained in the fuel (fuel NO) is oxidised. NO_x stands for the sum of nitrogen monoxide (NO) and nitrogen dioxide (NO_2).

Table 3.1. Shares of sectoral emissions of NO_x in the EU15 1994 (Source: CORINAIR 94)

SNAP	CORINAIR 90 Source Sector	Share
SNAP 7	Road transport	47.44 %
SNAP 1	Public Power, co-generation and district heating	18.95 %
SNAP 8	Other mobile sources and machinery	14.29 %
SNAP 3	Industrial Combustion	9.36 %
SNAP 2	Commercial, institutional and residential combustion	4.37 %
SNAP 4	Production Processes	1.70 %
SNAP 5	Extraction and distribution of fossil fuels	0.90 %
SNAP 9	Waste treatment and disposal	0.79 %
SNAP 10	Agriculture	0.24 %
SNAP 6	Solvent use	0.00 %
SNAP 11	Nature	1.96 %

4 Emission Abatement Measures

Demosthenes Papameletiou and Jose Maria Maqueda-Barrera

4.1 Introduction

The objective of this chapter is to identify and characterise measures for the abatement of NO_x and NMVOC, which are currently available and likely to become instrumental in the period until 2010.

As described in *Chap. 3* the sectors *road transport, fossil fuel combustion* and *solvent use* are the most important emission sources for ground-level ozone precursors. All together, these sectors are responsible for 66 % of NO_x and 58 % of NMVOC emissions at a European level (CORINAIR 94). Thus, the main emphasis for the identification of measures has been placed on these sectors.

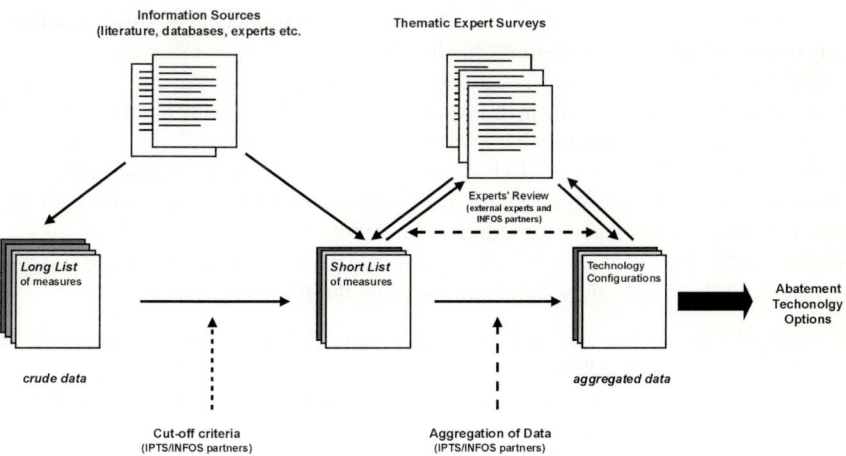

Fig. 4.1. Applied work approach and data model

A stepwise data collection approach was chosen to provide transparency in the process of selecting, storing, and sharing information. The structure of the applied data model was configured to enhance user friendliness and to allow comparability of the data, easy accessibility, interactive reviewing, and validation of key data for use in this study. The overall work approach is schematically depicted in *Fig. 4.1*.

First a *long-list* of emission sources of the pollutants, NO_x and NMVOC, and of emission reduction techniques was proposed on the basis of information provided by literature and feedback from external experts. In a second step, a *short-list* of emission reduction measures including the most relevant emission reduction technologies was selected with the aim of further collecting detailed data and of evaluating the emission reduction potential and their implementation costs. Finally, *thematic expert surveys* on fossil fuel combustion, road transport and solvent use industry were carried out to test and to validate the aggregation of the data into specific emission abatement measures, which represent currently best available techno-economically justified emission abatement prospects towards the year 2010. These are directly related to potential implementation solutions for existing and forthcoming regulations at European or national level.

4.2
Road Transport Sector

The road transport sector is one of the most important emission sources of ozone precursors pollutants. According to CORINAIR 94 about 47 % of NO_x and 31 % of NMVOC European emissions are due to road transport (*see Chap. 3*).

Although the emissions from a single road transport vehicle are rather small, the ever-increasing number of the circulating vehicles produces, in absolute terms, high quantities of these pollutants in the atmosphere. Since 1990 the European Commission has significantly toughened the regulatory framework for limiting ozone precursor emissions from transport vehicles. The already adopted *Euro 1* and *Euro 2* emission standards have been the first steps of the EU for limiting the emissions from passenger cars, light-duty and heavy-duty vehicles. New emission standards have been developed by the European Commission in the context of the Auto-Oil programme and were adopted in 1999. European associations of car manufacturers (ACEA[1]) and of oil industries (EUROPIA[2]) participated in this development, leading to new standards covering the following aspects:

(a) emission limits for cars and light duty vans (LDVs) to be implemented in 2000 (*Euro 3*) and 2005 (*Euro 4*),
(b) new specifications for gasoline and diesel limiting the content of sulphur, aromatics and benzene,
(c) the use of *on-board diagnostics* for emission monitoring, and

---

[1] Association des Constructeurs Européens d'Automobiles
[2] European Petroleum Industry Association

4.2 Road Transport Sector

(d) the adoption of a new test cycle for controlling emissions under *cold start* conditions.

Today, the Auto-Oil programme constitutes the most important data collection exercise at EU level. Consistently, the aim of the present study has been to take advantage of the Auto-Oil data and to generate - wherever possible - value-added. The Auto-Oil programme requested industry to provide information concerning technology options to meet pre-determined emission reduction modules/scenarios.

Starting from the results of Auto-Oil, the approach taken here was not limited by such given modules and therefore the determination of *best available performance data* was made possible and constitutes one major area of *value added* to the original Auto-Oil data. This study offers precise descriptions and characterisation of applied and emerging technology solutions towards the years 2005-2010, at a sufficiently dis-aggregated level by referring to published data. In particular, the new data up-date the current *state of the art* by focusing on developments emerging after 1995, which in several cases were not communicated by industry to the Auto-Oil programme. The overall results are shown in *Table 4.1*, which presents a comparative overview of the Auto-Oil data coverage on advanced motor vehicle technologies and emission abatement techniques and data used in this study. In addition, a prospective analysis was carried out in order to assess the potential of fuel cells. These have been selected as the most important low emission technology that could be applied in road transport before 2010.

The present study covers Passenger Cars (PCs), Light-Duty (LDVs) and Heavy-Duty-Vehicles (HDVs). Distinction is also made between gasoline and diesel fuelled vehicles. For each of these vehicle categories, various advanced system configurations were identified incorporating proposed abatement technologies expected to emerge up to the year 2010. The study illustrates expected improvements by comparing performances of the technology modules, the emission abatement measures with base cases technology modules representing the technology level available around the year 1990.

Emission factors representing the base case and the improved performances of advanced vehicles have been mainly obtained from published data. Publications by car manufacturers or catalyst developers are the principal sources of information on emission factors generated from testing results of emerging techniques. Information on the emission reduction economics in automobiles is extremely scarce. In this area, the Auto-Oil programme represents a unique information source, which provides industry evaluations about the cost of advanced automotive emission abatement measures.

Table 4.2 presents a description of the proposed technology modules as well as of the base cases. The base case for gasoline PCs is defined as a car technology before the introduction of three-way catalyst and without injection system. The base case for a diesel PC is a typical 1990 car operating without emission reduction aftertreatment systems and without improvement in the injection system. It has not implemented emission reduction aftertreatment systems and no improvement in the injection system were still made.

Table 4.1. Overview of Auto-Oil data coverage on advanced motor vehicle technologies and emission abatement techniques and data used in this study

	AUTO-OIL I data	data used in this study
Vehicle categories	Passenger cars (PCs) Light Duty Vehicles (LDVs) Heavy Duty Vehicles (HDVs) (gasoline and diesel fuelled)	Passenger cars (PCs) Light Duty Vehicles (LDVs) Heavy Duty Vehicles (HDVs) (gasoline and diesel fuelled)
Standard reference conditions for emission measurements [a]	ECE R15 + EUDC[b] test cycle (modified to include cold start emissions)	ECE R15 + EUDC test cycle NEDC-2000[c] ECE-R49[d]
Reference year for base case definitions	1996	1990
Description of base cases		
Technology	n.a.	Pre-1990 state of the art
Emission data	EU emissions standards 1996	Published data obtained at standard reference conditions
Economic data	n.a.	–
Description of proposed measures		
Technology	3 emerging technology modules/scenarios obtained through a car manufacturer survey carried out by Touche Ross (1995)	4 technology modules Synthesis of published data
Availability of technology	2000	1990-2005
Implementation period	2000-2010	1990-2010
Emission data	Expressed as achievable reduction levels in comparison to the base case (data does not represent best achievable performance levels)	Published data obtained at standard reference conditions (data represents best achievable levels)
Economic data	Data supplied by car manufacturers to Touche Ross (1995); Incremental costs related to the base case	Published data (1995-1998)

[a] ECE R15: Test cycle representing urban traffic conditions
[b] EUDC: Extra Urban Driving Cycle (road driving conditions)
[c] NEDC-2000: New European Driving Cycle that combine the above mentioned cycles with a modification (year 2000) for including cold start emissions
[d] ECE R49: Test cycle applied to heavy duty vehicles

Table 4.2. Advanced motor vehicle techniques selected as abatement options in this study (a) Passenger Cars (PC) and Light Duty Vehicles (LDV)

	PC and LDV Gasoline		PC and LDV Diesel	
	Availability	Technology configuration	Availability	Technology configuration
Base Case	PC and LDV: before 1990	PCs or LDVs without Three Way Catalyst (TWC) and injection systems	PC: before 1990 LDV: before 1992	1991 PCs or LDVs without any improvements in the injection system and without after-treatment technology
Technology Module 1	PC: 1992-93 LDV: 1993-94	TWC under-floor positioned Initial injection improvements	PC: 1992-93 LDV: 1993-94	Direct/Indirect injection Turbocharged injection
Technology Module 2	PC: 1996-97 LDV: 199-98	TWC under-floor positioned Injection optimisation: Dual Point Injection (DPI), Multi-Point Injection (MPI), Sequential Multi-Point Injection (SMPI)	PC: 1996-97 LDV: 1997-98	High speed Direct/Indirect injection Turbocharger and Intercooler Diesel oxidation catalyst (Pt/Pd) Exhaust gas recirculation Engine management system
Technology Module 3	PC and LDV: 2000	TWC under-floor positioned or close coupled positioned (Pt/Rh, Pd/Rh, Pt/Pd/Rh, Pd only) + Improved injection systems (MPI, SPI, SMPI) Thermal inertia of the exhaust system Exhaust gas recirculation Secondary air injection Engine management system	PC and LDV: 2000	High speed Direct/Indirect injection Electronically controlled turbocharger and intercooler Electronically controlled exhaust gas recirculation Advanced engine management system Diesel oxidation catalyst (Pt/Pd) De-NO_x catalyst (V/Ti/W, urea, Pt zeolite) Particulate traps
Technology Module 4	PC and LDV: 2005	TWC under-floor positioned (Pt/Pd/Rh, Pd/Rh) + Close couple positioned starter catalyst (Pd-only, Pt/Pd/Rh) or Electrically Heated Catalyst (Pt/Rh, Pt/Pd/Rh) + Advanced engine management system On Board Diagnostic (OBD) Secondary air injection	PC and LDV: 2005	Global improvement of the above mentioned technologies

Table 4.2. *cont.* (b) Heavy Duty Vehicles (HDV)

		Heavy-Duty Vehicles Diesel
	Availability	**Technology configuration**
Base Case	typical 1990 HDV	HDVs with small improvements of engine power and consumption (*retarding injection + Turbocharger + Intercooler*)
Technology Module 1	1992-93	Turbocharger Air-to-air or water-to-air Intercooler Redesigned injectors Higher injection pressures Variable injection timing
Technology Module 2	1995-96	Turbocharger, Intercooler Electrically controlled injection pump Variable injection timing strategy Engine management system Exhaust gas recirculation
Technology Module 3	2000	Advanced electronically controlled Turbocharger Advanced electronically controlled Intercooler Electronically controlled exhaust gas recirculation Engine management system Diesel oxidation catalyst (DOC) Catalytic trap oxidiser Advanced injection systems
Technology Module 4	2005	Previous configuration + De-NO$_x$ catalyst, Selective Catalytic Reduction (SCR), Particulate Traps

air/fuel ratio through cooling the Exhaust Gas Recirculation (EGR) in combination with electronic control of the combustion variables.

Particulate traps and diesel oxidation catalysts are the two available alternative options for reducing the emission of particulate matter. The particulate traps offer high performance efficiencies. However, there are no clear prospects about their future application because of problems encountered in the regeneration of the trapping material. An EMS could control this process but research results are still outstanding. Other problems related to the operation of the particulate traps include the increased fuel consumption, lower power output and safety considerations.

The combination of EGR with after-treatment processes, such as diesel oxidation catalyst offers a lower particulate reduction efficiency compared to the trapping systems. The advantage is that the need for a regeneration system is eliminated. In addition to the relatively low efficiency, another disadvantage is the potential for sulphate formation which promote the appearance of abrasive and corrosive wears by particles and by sulphur respectively. Pd based catalysts are preferable for heavy-duty vehicles, because of the higher temperature of the exhaust gases. Additionally the sulphate formation potential is lower.

Finally selective catalytic reduction and non-selective catalytic reduction are also under discussion as options for NO_x emission reduction. Continuous supply of the reducing agent (urea/ammonia) and the emissions of urea slips are the main difficulties. The use of unburned hydrocarbons as reducing agents has also been proposed. This could be achieved either through addition of secondary fuel to the exhaust stream or through deterioration, on purpose, of the combustion process.

4.2.4
Fuel Cell Powered Vehicles

Fuel cells offer the advantage of highest fuel consumption efficiencies at almost zero emissions. During the last three decades they have been promising to substitute the internal combustion engine but these prospects started to approach the commercial breakthrough only during the last few years. From the several types of fuel cells currently under demonstration in various application areas, only the *proton exchange membrane* offers the necessary operational and power density characteristics for automotive applications (small and light, low operating temperature, and quick response time to frequent vehicle variable dynamics). During the current decade strong efforts and high investments were made by fuel cell companies, car manufacturers and governments with focus on basic research and on the integration of the different parts of the fuel cell system (fuel cell stack, fuel processor and power section). Following several early demonstration cases, DaimlerChrysler offered the first short-term prospects for *a commercial penetration in 2004* in a recent announcement. This announcement is based on the results obtained from the demonstration of the last fuel cell powered prototype, the 4th version of the NECAR (New Electric Car) in March 1999. This vehicle is based on a Mercedes-Benz class A compact model and is fuelled by liquid hydrogen stored in a cryogenic cylinder. The high level of system integration regulation of the injection process depending on the operation conditions maintaining the engine efficiency at a high level, without fuel consumption penalties. Novel injection systems can produce an optimal fuel droplet size and distribution through the increased injection pressure. Furthermore, turbochargers and inter-coolers of the intake air permit an increase in the air fuel ratio and thus improve the overall efficiency of the engine.

The use of oxidation catalysts for trapping particulates provokes a trade off regarding the reduction efficiencies for NO_x and for particulates. This calls for a careful optimisation of the overall system. The introduction of EMS enables the application of flue gas recirculation techniques for NO_x abatement. Further going abatement can only be accomplished through catalytic reduction of NO_x. Intensive research activity has contributed to the development of De-NO_x catalysts for application in diesel cars. Platinum and cooper zeolite catalysts are currently the prevailing choices. The reducing agent consists of unburned hydrocarbons, the supply of which is ensured either through the deterioration of combustion or through hydrocarbon addition to the exhaust gas

The problem of particulate matter is the most difficult to tackle, in particular regarding the emission reduction of the smaller fraction which causes the main threats for human health. Current emission standards can be met by existing technologies. However, the emission of the smaller fraction still represents an important threat, which justifies further technology development and regulatory initiatives.

4.2.3
Heavy-Duty Vehicles – Diesel

The NO_x, HC and most of the particulate emissions from diesel fuelled vehicles are formed during the combustion process and can be controlled by appropriate modifications of the combustion process. The *Euro 2* emission standard can be met by engine modifications and by improvements in the injection systems. Stricter emission standards (*Euro 3* to be adopted by 2000) will require after-treatment of the exhaust gases, in addition to combustion modifications.

In recent years, better understanding of diesel combustion processes and of factors affecting pollutant formation has been the key issue for improving engine designs. Of particular importance are the so-called in-cylinder techniques, such as modifications of the combustion chamber geometry and variable air swirl in order to achieve best mixing conditions under different speed/load ranges. Variable pressure and geometry turbochargers were designed depending on loads and speed levels. High efficiency air-to-air or air-water inter-coolers were necessary in order to reduce the compressed air temperature as much as possible. NO_x emissions can be reduced with fuel injection timing retard, while greater injection pressures (rotary injection pumps, common rail and unit injectors) reduce the formation of particulate matter. The introduction of Engine Management Systems (EMS) has permitted the control of all these processes through electronics, optimising combustion and injection variables for the different speed/load ranges. The generation of particulate matter emissions can be decreased by optimising the which is called close-coupled position. Catalysts in this position need to withstand higher temperatures and suffer greater mechanical stress through vibration. In this respect several problems related with the thermal durability and mechanical stability of the catalyst still need to be resolved.

Future and most stringent emission standards for cold start emissions (*Euro 4* for 2005) will most likely impose the adoption of some kind of heating systems applied to the catalysts. According to available test results, electrically heated catalytic converters seem to offer better performance than fuel-burner systems. This technique is already in the development stage, according to industry reports considerable efforts and time are still needed for the achievement of commercial goals.

The development of electronics permits the real time monitoring of the emissions and the diagnosis of engine performance. The Engine Management System (EMS), electronically controlled injection and the On Board Diagnostic (OBD) system are some of the techniques currently under demonstration. Depending on the software incorporated, several functions and techniques can be combined to lower emissions and to increase the overall system efficiency. These techniques include the internal exhaust gas recirculation, secondary air injection, late ignition, and mixture enrichment in order to meet requirements for specific situations such as cold starts and variable loads.

4.2.2
Passenger Cars and Light-Duty Vehicles – Diesel

Diesel fuelled engines are inherently a low emission technology option producing less CO emissions compared to the gasoline fuelled engines. Particulate and NO_x emissions are the main environmental problems.

The carcinogenic effects of particulate matter can be prevented with oxidation catalysts, which are widely used today in order to reduce the soluble organic fraction (SOF) of particulate emissions. Diesel catalytic converters oxidise a large part of the SOF of particulates, as well as gaseous HC, CO and odour. The basic formulations use Pt and Pd as active materials properly deposited on ceramic or metallic supports. Potential formation of sulphate can occur, which additionally increases particulate emissions. Reformulated fuels with lower sulphur content could avoid this problem.

Current research activities are devoted to increasing the trapping capacity of the catalysts. The addition of zeolites (also known as molecular sieves) to the catalyst formulation promotes the trapping capacity of the catalyst during low engine load or idle period before the catalyst reaches its peak operation temperature. In addition to oxidation catalysts, purely trapping systems have been widely and successfully applied in heavy-duty vehicles. The main problem is the regeneration of the trap. Different technologies are being developed and applied using electrical heating or additives for regeneration.

Optimisation systems for the combustion and injection sections have been developed in parallel to the development of Engine Management Systems (EMS). Retarded injection timing electronically controlled is intended to permit the

A representative 1990-93 LDV has been taken as the base case for the implementation of abatement technologies and for the calculation of emission reduction achievable with the proposed technologies. A naturally aspirated HDV engine with retard of injection timing for NO_x abatement has been selected as a reference (although turbocharger and inter-cooler began to be introduced).

Emission factors measured under standardised test cycles prescribed by the European and American[3] legislation were used. This permitted comparability among data selected from different information sources. The most advanced version of the New European Driving Cycle (NEDC) was selected as a standard reference for measures for PCs and LDVs. This test cycle is representative for urban and highway driving conditions. It does not cover "cold-start" emissions, as it is the case with the NEDC-2000 driving cycle for which very few test results have been published. Finally, the ECE-R49 test was selected as a standard reference for HDV data.

4.2.1
Passenger Cars and Light-Duty Vehicles – Gasoline

Currently, there are different emission standards applied to passenger cars and to light-duty vehicles. However, in most of the cases the technological improvements for PCs can also be implemented in LDVs, as the engine design characteristics tend to be similar. Compliance with the most stringent emission standards proposed for the year 2005 clearly requires combination of several solutions including improved engine techniques, after-treatment processes, as well as better fuel qualities. Considerable advances are being reported, but in many cases car manufacturers claim that the available time period is too tight for the full development of several of these solutions. The introduction of Three-Way Catalyst (TWC) in under-floor position early in the 90's and several improvements in fuel combustion thanks to novel injection techniques (dual point injection, multi-point injection and sequential multi-point injection) are the main techniques currently implemented. Furthermore, of key importance is the evolution in catalyst formulations starting from the early Pt or Pt/Rh based concepts.

Today, the most advanced formulations containing Pd or combination of Pt/Pd/Rh are the basis for complying with the proposed *Euro 3* version of the European standards. Increasing the palladium content in catalysts is an option currently under investigation for achieving standards going beyond this level. This is technically achievable but the cost issue seems to be an inhibiting factor for a large scale application. Novel system designs, especially regarding the positioning of the catalyst allow the achievement of cost advantages with rather conventional catalyst formulations. The control of the "cold start" emissions requires the application of techniques, which permit a quick activation (light-off) of the catalyst. This is achievable either by improving the thermal inertia of the exhaust system or by installing the catalyst in a position as close as possible to the engine,

---

[3] Data obtained under the American Federal Test Procedure (FTP) were converted to European equivalents by using appropriate correlation and conversion factors.

emissions. NO_x and NMVOC emissions from the stationary combustion of fossil fuels are mainly generated in power and heat production processes. Public power, industrial combustion and commercial, institutional and residential combustion are the emission sources taken into consideration. During the past decade, the systematic installation of emission reduction equipment has been a main result of the implementation of the Large Combustion Plant (LCP) Directive[4] and of national regulations in almost all the European countries. The main pollutants covered are particulates, NO_x and SO_2. Regarding the setting of emission standards for NO_x, several Member States have been playing a pioneering role. The most representative is the case of Germany, where stricter values compared to those adopted at EU level were introduced at an early phase. This led to significant technology development efforts and demonstration results, presently available to the rest of the EU. In this frame, the proposed revision of the LCP directive[5] as well as the development of a concept for a "National Emissions Ceilings" (NECs) directive for both new and existing plants, are opening a window for the adoption of currently achievable low level emission limits at a Pan-European scale. The potential for further NO_x emission reduction in the EU can be approximated from information regarding the percentage of power plant capacity that needs to be equipped with de-NO_x abatement measures (*Table 4.5*).

Two types of techniques are principally applied for NO_x abatement: *primary measures* (or combustion modifications, CM) and secondary measures (or *flue gas treatment processes,* FGT). Distinction between retrofitting and new plants is usually made to highlight the substantial technical and economic constraints encountered in the case of retrofitting. *Low excess air, over fire air, flue gas recirculation, low NO_x burners, fuel reburning, fuel staging, staged air combustion* are typical combustion modifications applied world-wide in power generation. Progress made in combustion monitoring has permitted the introduction of operational controls for optimising these techniques. In general, combustion modification techniques are usually applied in several possible combinations in power plants. However, for the purpose of this study, these measures were combined into one aggregated package.

Furthermore, two types of FGT processes were selected for further investigation: *Selective Catalytic Reduction* (SCR) and *Selective Non-Catalytic Reduction* (SNCR). Both have been successfully applied at a large scale in Germany and represent the most important options for compliance with emerging emission standards.

Primary measures are widely applied in European utility boilers. However, it is conceivable that the introduction of the proposed revision of the LCP Directive will require the adoption of FGT processes to achieve higher emission reduction levels, however, at high costs. According to the information presented in *Table 4.5*, in Germany almost 55 % of the power generating capacity has been equipped with SCR abatement techniques, while the level of implementation of

---

[4] Council Directive 88/609/EEC, on the limitation of certain pollutants into the air from large combustion plants. OJ L 336, 7.12.88
[5] Proposal for a Council Directive amending Directive 88/609/EEC, COM (98) 415 final. OJ C 300/98.

primary measures is around 14 %. In Austria the situation is quite different with 54 % of implementation degree of combustion modifications and 35 % of SCR process. On the other hand, Mediterranean countries such as Spain, Italy and Greece need to make strong efforts to comply with the proposed directive due to the poor application of de-NO_x measures and the high investment required.

Combustion modifications are extensively used for *institutional, commercial and residential combustion*. Air combustion staging, low NO_x burners (LNB) and flue gas recirculation are the most widely implemented techniques, with the support of combustion monitoring. Although FGT could also be implemented, space availability as well as highest relative costs when compared with combustion modifications are the main barriers for the application of these secondary abatement measures.

Finally emission reduction can be achieved by adopting cleaner combustion techniques but the commercialisation prospects for such techniques are not yet clear. Advanced power generation technologies, such as fluidised bed combustion or integrated gasification combined cycle were not considered as technology options for broad application towards the year 2010. Higher investment costs and quality limitations regarding the operational fuels are the main barriers.

4.3.1
Techno – Economic Profile of Measures for the Energy Sector

New plants are currently standard designed to incorporate combustion modification measures as described above. Retrofitting a boiler with primary measures may be difficult in terms of existing space, boiler size and type, re-burning fuel availability and global emission reduction reached. Furthermore every measure can be accompanied by undesirable effects (unburned carbon, slagging and furnace corrosion) that need to be taken into consideration. SCR can be applied to flue gases of diesel engines, flue gases from oil, coal, and natural gas fired boilers, municipal solid waste incinerators, and for flue gases from industrial processes such iron, glass and cement production. It seems to be the best option for the achievement of future emission limits, despite minor reported problems related with ammonia slips in flue gases. SNCR can be applied to coal, oil and natural gas fired installations. This process is well suited for plants with low NO_x content in flue gas. It is easier to operate than a SCR system as it does not use a catalyst but it requires maintenance of the temperature inside a given "temperature window" in order to achieve the maximum performance.

Table 4.5. Fossil Fuelled Power Plants in Europe – Installed NOx Abatement Capacities (reference year: 1995)

EU Countries	Total Generating Capacity (GW)	Total Power Plant Capacity (GW)			COAL				OIL				NATURAL GAS			
		Coal	Oil	Natural Gas	Primary Measures (GW)	Secondary Measures[c] (GW)		Primary Measures (GW)	Secondary Measures (GW)		Primary Measures (GW)	Secondary Measures (GW)	
						SCR[a]	SNCR[b]		SCR	SNCR		SCR	SNCR
Austria	17.87	1.89	0.71	4.68	1.67	1.26	0.41	0.16	0.65	0.23	2.12	0.65	0.00
Belgium	15.57	4.85	0.58	2.56	4.80	0.00	0.00	0.35	0.00	0.00	2.17	0.00	0.00
Denmark	10.67	6.43	1.16	2.00	4.19	1.14	0.04	0.15	0.00	0.00	0.41	0.00	0.00
Finland	16.50	5.39	1.61	2.55	1.81	0.56	n.a.	n.a.	n.a.	n.a.	n.a.	n.a.	n.a.
France	115.01	11.20	11.70	1.85	0.00	0.00	n.a.	n.a.	n.a.	n.a.	n.a.	n.a.	n.a.
Germany	115.50	55.10	8.70	18.60	10.00	36.70[d]	2.00[d]	0.00	2.23	0.00	0.00	0.93	n.a.
Greece	11.80	4.83	2.28	1.41	n.a.	n.a.	n.a.	n.a.	n.a.	n.a.	n.a.	n.a.	n.a.
Ireland	4.66	1.26	0.82	1.87	0.31	n.a.	n.a.	n.a.	n.a.	n.a.	n.a.	n.a.	n.a.
Italy	67.55	9.12	18.64	17.57	3.49	3.49	n.a.	n.a.	n.a.	n.a.	n.a.	n.a.	n.a.
Luxembourg	1.26	0.09	0.00	0.02	0.00	0.00	n.a.	n.a.	n.a.	n.a.	n.a.	n.a.	n.a.
Netherlands	19.23	3.92	0.01	13.50	4.30	1.24	n.a.	0.00	0.00	0.00	n.a.	n.a.	n.a.
Portugal	8.41	1.87	1.92	1.32	1.87	0.00	0.00	0.00	0.00	0.00	0.99	0.00	0.00
Spain	50.68	11.74	5.38	5.05	4.29	0.00	0.00	0.32	0.00	0.00	2.05	0.00	0.00
Sweden	33.62	0.70	5.23	0.32	0.73	0.25	n.a.	n.a.	n.a.	n.a.	n.a.	n.a.	n.a.
United Kingdom	72.21	34.23	8.42	10.16	12.60	0.27	0.26	n.a.	n.a.	n.a.	n.a.	n.a.	n.a.

[a] SCR: Selective Catalytic Reduction
[b] SNCR: Selective Non-Catalytic Reduction
[c] or combination of primary and secondary measures
[d] almost all large combustion plants in West Germany have been equipped with secondary measures, plants in the former GDR had to comply by 1996

Source: INFOS Thematic Expert Survey "Power Sector"

Table 4.6. Technology measures for emission reduction in the energy production sector: *Large Combustion Plants* (LCP) and *institutional and commercial combustion plants*

Plant-Category	Measure	Fuel type	Category[1]	Life time	EF NO$_x$[2] before	NO$_x$ emission reduction	Investment (ECU/kW$_{el}$)	O&M[3] (ECU/kW$_{el}$)	Labour (ECU/kWh$_{el}$)
Large Combustion Plants	Primary Measures	Coal	R	30	400	60%	44	2.64	-
			N				33.88	2.0328	-
		Oil	R	30	180	66%	28	1.68	-
			N				21.53	1.2918	-
		Gas	R	30	140	92%	28	1.68	-
			N				21.53	1.2918	-
	Selective Catalytic Reduction (SCR)	Coal	R	20	400	80%	70	8.51	0.001
			N				37	16.1	0.001
		Oil	R	20	180	80%	45	5.75	0.001
			N				25	10.35	0.001
		Gas	R	20	140	80%	50	4.6	-
			N				20	11.5	-
	Selective Non-Catalytic Reduction (SNCR)	Coal	R	20	400	50%	15	7.5	-
			N				10	11.25	-
		Oil	R	20	180	50%	15	7.5	-
			N				10	11.25	-
		Gas	R	20	140	50%	12	6	-
			N				8	9	-
	Hybrid SCR/SNCR	Coal	R	20	400	70%	36	11.55	-
			N				21	19.8	-
		Oil	R	20	180	70%	36	11.55	-
			N				21	19.8	-
		Gas	R	20	140	70%	36	11.55	-
			N				21	19.8	-
Institutional & Commercial Combustion	Primary Measures	Coal	R	30	230	50%	13.9	0.83	-
			N				10.7	0.64	-
		Oil	R	30	120	60%	8.9	0.53	-
			N				6.9	0.42	-
		Gas	R	30	85	60%	8.9	0.53	-
			N				6.9	0.42	-

[1] R/N: Retrofitting/New
[2] EF: Typical emission factor before the application of abatement techniques
[3] O&M: Operating and Maintenance costs

Source: INFOS Thematic Expert Survey on the Energy Sector

Table 4.6 shows the aggregation level of the control technology, technical and economic performance characteristics selected as representative for currently available techniques for NO_x abatement. Typical *uncontrolled* emission factors and average emission reduction levels are provided in addition to the costs of emission abatement equipment for existing and new installations. For the purpose of this study, no distinction by boiler size was necessary, mainly due to relatively minor cost differences. The uncontrolled emission factors were adopted from CORINAIR-90 data, except for small combustion where only U.S. EPA data were available. In general, the achievable emission reduction level through application techniques depends on the design of the plant, the operating regime and the fuel specification. In the case of the selected abatement measures, these levels were selected in such a way to represent typical operational performances of *best practice* plants in the EU.

Indicative cost estimations for the investment and operating costs of de-NO_x measures were directly provided by industry and seem to straightforward represent typical European conditions. Relatively low investment is needed in the case of combustion modifications, as opposed to alternative options for flue gas treatment. SCR costs are determined by process position (high/low dust), space availability in the case of retrofit projects, operation temperature, catalyst wear, and first catalyst filling. SNCR operates without catalyst and offers favourable economics compared to SCR in cases where relatively low reduction levels are acceptable.

Long years development efforts and operating experiences have contributed to the consolidation in technical and economic terms of several standard solutions in this area. In this light, it is likely that the application across the EU of most of these currently available solutions will be the prevailing investment options for the abatement of NO_x emissions within the time horizon of the present study.

4.4
Solvent Use Sector

This sector comprises industries using any kind of solvent products in their production processes. According to the inventory of atmospheric emissions in the EU (CORINAIR 90) the contribution of the solvent user industries to the total NMVOC emissions in Europe is in the range of 29 %. It is the second most important emission source following the road transport sector, which contributes to 42 % of the total emissions *(see Chap. 3)*.

The recently adopted European solvent directive[6] provides the basis for regulating the emissions reduction in this sector. It sets emission standards for almost twenty types of solvent use applications. One of the key proposals refers to the establishment of a monitoring network in European countries, which needs to be developed as a necessary support measure for the control of NMVOC emission levels.

---

[6] Council Directive 1999/13/EC on the limitation of emissions of volatile organic compounds due to the use of organic solvents in certain activities and installations, OJ L 85/1, 29.03.1999

4 Emission Abatement Measures

Table 4.7. NMVOC emitting sectors and related activities covered by this study

Solvent sector	Activity	Description
Surface coatings	furniture coating	any activity related with the industrial coating of wooden surfaces
	coil coating	any activity where coiled steel, stainless steel, coated steel, copper alloys or aluminium strip is coated with either a film or forming or laminated coating in a continuous process.
	film coating	any activity in which a single or multiple application of a continuous film of a coating is applied to textile, fabric, film and paper surfaces
Surface cleaning	metal degreasing	any activity except dry cleaning using organic solvents to remove contamination from surface of material (products not equipment) including degreasing
Printing	publishing, flexible packaging, decoration	any reproduction activity of text and/or image in which ink is transferred onto whatever type of surface. It includes publication, packaging and decoration activities. The approved directive covers the following processes widely used by the printing industry: lithography, rotogravure, flexography, and screen printing, laminating and varnishing
Industrial use of paints	vehicle manufacture	any activity in which a single or multiple application of a continuous film of a coating is applied to a vehicle, as part of the manufacture process
	vehicle refinishing	any industrial or commercial coating activity and associated degreasing activities performing the coating of road vehicles or part of them, carried out as part of vehicle repair, conservation or decoration outside of manufacturing installations
Rubber processing	general rubber goods, tyre manufacture	any activity of mixing, milling, blending, calendering, extrusion and vulcanisation of natural or synthetic rubber and any ancillary operations for converting natural or synthetic rubber into a finished product
Industrial adhesives	adhesive and sealant use	any activity in which an adhesive is applied to a surface, with the exception of adhesive coating and laminating associated with printing activities

The extreme diversity of industrial processes, which use solvents and generate NMVOC emissions, represents an important difficulty in developing uniform and efficient regulations for controlling and monitoring this type of emissions. Depending on the type of utilisation of the solvent raw materials, these emission sources can be classified either as industrial or domestic sources. Applied

The data on the techno-economic assessment of available alternative emission abatement options for implementation in the road transport sector in the period 1990-2010 is presented in *Table 4.3*.

Availability and the date of the commercial introduction of the proposed technology modules is presented to highlight potential correlation with the existing and forthcoming emission regulations as follows: *module 1* with Euro 1, *module 2* with Euro 2, *module 3* with Euro 3, and *module 4* with Euro 4. The emission factors reported in each case correspond to best achievable performance characteristics and indicate the potential for compliance with emission standards. The expectable emission reduction can be calculated by combining the performance data for different measures with the corresponding data of the base cases.

In addition to the above mentioned *conventional* measures, fuel cells can also be considered as a separate, (almost) zero emission measure for the road transport sector. A potential market development seems to be a near-term prospect for fuel cell powered vehicles in the urban transportation sector. *Table 4.4* presents early prospective evaluations of the cell stack and the system costs for several fuel options.

Table 4.4. Efficiency and Cost estimation of Automotive PEM[a] Fuel Cell Systems

Fuel	Year	Efficiency (%)	Stack[b] cost (ECU/kW)	System[c] cost (ECU/kW)
Oil	1994	30 – 50	90 655	90 655
	2000	35 – 55	9 065	18 135
	2010	36 – 60	905	1 815
Methanol	1994	38 – 45	90 655	-
	2000	40 – 50	6 350 – 9 065	2 720 – 22 665
	2010	42 – 52	455 – 905	905 – 2 720
Natural gas	1994	40	90 655	-
	2000	45	9 065	13 600
	2010	50	905	4 535

[a] PEM: Proton exchange membrane
[b] Stack refers to the group of cells where electrochemical reaction is produced
[c] System refers to the complete fuel cell system, including fuel processor and power conditioner section

Source: D. Papameletiou: Fuel Cells, Report EUR 16321 EN, IPTS, 1994

4.3 Energy Sector

According to the inventory of atmospheric emissions in the EU (CORINAIR 94) stationary fossil fuel combustion generates almost 19 % of the total NO_x emissions in the EU. NMVOC emissions from this sector account for only 1 % of the overall achieved will permit the utilisation of other fuels in the future. Although hydrogen is considered to be a good option for heavy-duty vehicles or busses, its application for passenger vehicles will be limited by the infrastructure necessary for refuelling operations. This is a key issue to be resolved, in particular to ensure public acceptance and to minimise the necessary changes to the existing infrastructure.

4.2.5
Techno-Economic Profiles of Measures for Road Transport Sources

Table 4.3. Abatement options for the road transport sector

Vehicle Category	Fuel	Techn. Modules[1]	Technology Availability	EF_{NOx} (g/km)[2]	EF_{NMVOC} (g/km)	Investment (ECU/veh)	Labour and O&M (ECU/veh)
Passenger Cars	Gasoline	Base Case	before 1990	2.50-2.80	1.57-2.04	-	-
		Module 1	1992-1993	0.42-0.69	0.13-0.33	335	25
		Module 2	1996-1997	0.14-0.27	0.13-0.23	335	25
		Module 3	2000	0.03-0.15	0.07-0.18	577	62
		Module 4	2005	0.03-0.09	0.03-0.08	839	86
	Diesel	Base Case	before 1990	0.830	0.210	-	-
		Module 1	1992-1993	0.446	0.095	145	-
		Module 2	1996-1997	0.43-0.72	0.05-0.09	145	-
		Module 3	2000	0.357	0.045	769	-
		Module 4	2005	0.214	0.032	919	-
Light-Duty Vehicles	Gasoline	Base Case	before 1993	2.710	1.360	-	-
		Module 1	1993-1994	0.470	0.169	335	25
		Module 2	1997-1998	0.282	0.143	335	25
		Module 3	2000	0.141	0.072	660	65
		Module 4	2005	0.100	0.050	961	90
	Diesel	Base Case		1.420	0.650	-	-
		Module 1	1993-1994	1.358	0.197	145	-
		Module 2	1997-1998	1.251	0.155	145	-
		Module 3	2000	2.420	0.078	527	-
		Module 4	2005	2.380	0.015	708	-
Heavy-Duty Vehicles	Diesel	Base Case	before 1990	7.5-11.0	0.48-1.35	465	-
		Module 1	1992-1993	5.0-10.0	0.2-0.6	700	-
		Module 2	1995-1996	4.7-7.6	0.12-0.49	1 405	-
		Module 3	2000	2.33-5.87	0.04-0.07	2 920	-
		Module 4	2005	1.5-3.0	0.012-0.18	5 939	-

[1] Module 1 to Module 4 represent technology configurations described in Table 4.2
[2] EFs= Emission factors

Source: INFOS Thematic Expert Survey "Road Transport Sector".

emission reduction practices in industrial applications are based on solvent management plans, on modifications of the processes and/or on the application of gas treatment techniques. In the case of the domestic use of solvent, emission reduction can only be achieved with the substitution of traditional solvent-based products with newer less polluting formulations.

Table 4.7 presents the industrial processes covered in this study. The selection of the most important solvent use sectors was made on the basis of information from recent studies carried out on behalf of the European Commission, which have been evaluating the economic implications of the solvent Directive. Regarding the domestic use of solvents as household and personal care products, adhesives and glues, automotive maintenance products as well as non-industrial use of paints, there is a pronounced scarcity of data. In these areas the most important techniques include the modification of the application technique (from over-spray to electrostatic spray, but not fully implemented) and the solvent substitution (water-based, high-solids and powder coatings). Country specific parameters regarding the application of these techniques are currently surveyed by dedicated studies on behalf of the European Commission.

4.4.1 Techno–Economic Profiles of Measures for the Solvent Sector

A wide range of emission abatement techniques is currently available for reducing VOC from stationary sources. These can be grouped into the following four categories:

- **Good housekeeping practices.** In general terms, good housekeeping consist of a number of measures that allow a better exploitation of existing resources. The application of such practices does not require excessive amount of investment and the achievable emission reduction is relatively low. However, in several cases the emission reduction potential achievable by good housekeeping can lead to compliance with regulation. In this context, solvent management plans, as well as leak monitoring and repair programs are the most widely used techniques for emission reduction. Specific training programs for operators in the techniques of handling solvents or in applying them to a surface in an efficient way, by using correctly the equipment, are also management options to be applied.
- **Substitution** of substances with a high solvent content by other products which are environmentally less harmful. The aim of this technique is the reduction of solvent consumption. Depending on the activity, local specifications for certain products are often a major difficulty for the adoption of substitution measures. The current substitution trend in painting and coating processes, is towards water-based, high-solids and powder coatings. Substitution measures need often to be combined with modifications on the application methods. For some sectors, such as rubber-processing and surface cleaning, research activities are still needed regarding the improvement of solvent substitute formulations in order to meet specific safety and quality requirements.
- **Process modifications** offer frequently the best opportunity to increase the process efficiency and to reduce solvent consumption. In many cases, the

reduction in the solvent consumption results as a sufficient measure to meet emission standards. The applicable process modifications are numerous and their economics widely scattered and strongly depending on the nature of the original processes.
- **End-of-pipe** control options are generally considered as efficient but high cost solutions. Thermal or catalytic incineration and carbon adsorption for solvent recovery are the most widely applied technologies. In addition, there are other techniques such as absorption, wet scrubbers, bio-treatment and carbon canisters for solvent recovery, that can be applied in specific cases.

Table 4.8. Abatement measures for the solvent sector

Industry Sector	Emission Control Techniques	NMVOC emission reduced	Applicable to [h]	Investment (ECU/(t/yr))	O&M (ECU/t)	Savings (ECU/t)
Furniture Coating	Good housekeeping	10%		0.0	0.0	n.a.
	Substitution [a]	65%		6806.1	3880.5	
	Process modification [b]	15%		0.0	n.a.	
	Thermal oxidation and adsorption	70%		9737.1	375.7	
Coil Coating	Good housekeeping	2%		0.0	0.0	n.a.
	Thermal oxidation	93%		1050.6	128.1	
Film Coating [c]	Thermal oxidation	84%		1608.4	160.1	n.a.
Vehicle Refinishing	Good housekeeping	5%		0.0	0.0	0.0
	High Volume Low Pressure (HVLP)	13%		1577.8	0.0	7846.4
	Enclosed gun-wash	26%		1628.7	0.0	194.3
	HVLP + Substitution (high solids)	28%		814.3	2595.7	6273.0
	HVLP + Substitution (water borne)	44%		2850.2	4326.2	0.0
Vehicle Manufacture	Good housekeeping	30%		0.0	0.0	n.a.
	Substitution (medium solids)	25%		0.0	1215.0	
	Substitution (water based)	30%		4732.7	874.7	
	Thermal oxidation	35%		3810.0	291.6	

4.4 Solvent Use Sector

Table 4.8. Abatement measures for the solvent sector, *continued*

Surface Cleaning	Good housekeeping	15%	67%	283.8	0.0	252.7
	Substitution [d]	100%	45%	19908.7	3427.8	0.0
	Improved design	33%	100%	3951.6	0.0	268.2
	New enclosed system	70%	4%	14858.4	0.0	1107.0
	Single sealed chamber	85%	1%	74390.4	0.0	1230.0
	Double Lidded System (DLS)	70%	18%	5147.2	481.2	481.2
	DLS with carbon adsorption	85%	19%	21061.7	598.8	810.0
Printing Industry	Good housekeeping	5%	70%	0.0	0.0	n.a.
	Thermal oxidation (publishing)	93%	90%	3254.7	124.0	n.a.
	Thermal oxidation (flexible packaging)	78%	80%	1670.2	65.3	n.a.
	Thermal oxidation (decoration)	78%	80%	4495.2	294.3	n.a.
General Rubber Goods	Housekeeping	13%	63%	61.9	49.5	61.9
	Substitution [e]	100%	25%	142.7	21.4	0.0
	Process modification [f]	13%	33%	123.8	45.9	16.0
	Thermal oxidation	74%	70%	2166.9	215.5	0.0
	Carbon adsorption	74%	20%	4100.5	494.8	156.7
Tyre Manufacturers	Good housekeeping	13%	63%	36.9	3.7	36.9
	Substitution [e]	100%	25%	135.3	20.3	0.0
	Process modification [f]	13%	33%	73.8	0.0	36.9
	Thermal oxidation	74%	70%	412.7	36.9	0.0
	Carbon adsorption	74%	20%	766.3	143.9	97.2
Adhesive and Sealant Use	Good housekeeping	15%	70%	0.0	0.0	n.a.
	Substitution [g]	85%	60%	318.9	139.3	n.a.
	Process modification	13%	70%	0.0	0.0	n.a.
	Thermal oxidation	78%	45%	837.2	93.0	n.a.

[a] Substitution of traditional solvent paints by high-solids and waterborne paints
[b] Air assisted/airless application, curtain/roller coating and dipping, HVLP (high volume low pressure) spray guns and other more efficient techniques
[c] Masking film, drawing office film, foil printers, …
[d] Substitution of traditional solvents by hydrocarbon semi-aqueous or water based solvents
[e] High solids and water based products that meet safety, quality and end-user requirements
[f] Improvements on the solvent delivery system and on the production lines minimising the amount of solvent and the refreshing requirements.
[g] High solids, water based and hot melt adhesives
[h] referring to the share of plants operating in the EU that can be covered by this measure
All given costs are expressed per tonne solvent consumed

Source: INFOS Thematic Expert Survey of the Solvent Use Sector

Table 4.8 shows aggregated information on control technology and technical and economic performance characteristics selected as a basic input to the selection of measures for NMVOC abatement in the solvent use sector. The existing potential for emission reduction through application of these measures is substantial. According to preliminary evaluation results regarding the implementation of the solvent directive, total NMVOC emission reduction in the solvent sector across the EU could reach by 2007 a relatively high level in the range of 50 – 60 %.

4.5
Conclusions

The aim of this chapter was to describe the current state of development of abatement techniques for the ozone precursors NO_x and NMVOC, their techno-economic characteristics and the prospects for the implementation of measures towards the year 2010.

The road transport sector is considered as the most important source for the emission of ozone precursors in the EU. A wide variety of advanced techniques are already available. However, emission standards tend to become more and more restrictive in Europe and in the United States and provide for the next decade a highly dynamic framework for regulation driven innovations. In this light, it is conceivable that by 2010 the zero emission car will be a reality. Fuel cells represent the most promising breakthrough in this area. Almost emission-free conventional cars are also within the reach of commercial availability within the time period under consideration. In this respect, the outcome of current debates in the EU will determine the pace of further technology development and implementation, in particular for heavy-duty vehicles.

Emission reduction technologies needed in the sector of *fossil fuel combustion* for power generation already exist and have been widely and successfully demonstrated in several pioneering Member States of the EU. For this sector the adoption of legislation in the EU is running along behind the technology development. However, the installation degree of commercially available abatement measures is uneven across Europe. In this light, the key issue during the coming decade will be the pan-European implementation of the forthcoming Large Combustion Plants Directive, which is currently under revision.

From the point of view of technology availability, the *solvent use sector* represents an intermediate case between the above two sectors. On one hand *end of pipe* solutions are fully developed and widely applied. On the other hand, emerging solutions such as process modifications, material substitution and the introduction of clean technology can significantly contribute to emission reduction at economically favourable conditions compared to the application of end of pipe techniques. For both applications – end of pipe techniques as well as cleaner technologies – a sound amount of commercial solutions are already available. However, the uneven implantation within the EU of emission abatement measures in the solvent user industries needs to be tackled during the next decade under the pressure of the on-going implementation of the solvent Directive. During this decade, it is conceivable that innovation will be on the side of clean technology solutions, which will improve economics and system performance, thus lowering the demand for end of pipe techniques.

4.6 References

Road Transport sector

Rodt S, Friedrich A et al (1996) HDV-2000 – Requirements, Technical Feasibility and Costs of Exhaust Emission Standards for Heavy-Duty Vehicle Engines for the year 2000 in the European Community. Federal Environment Agency, Berlin

Rodt S, Friedrich A et al (1995) Passenger cars 2000 – Requirements, Technical Feasibility and Costs of Exhaust Emission Standards for the year 2000 in the European Community. Federal Environmental Agency, Berlin

OECD (1993) – Control of Emissions from Heavy-Duty Vehicles.

Air Quality Report of the Auto-Oil Programme. Report by the Directorate Generals for Industry; Energy; and Environment, Civil Protection & Nuclear Safety of the European Commission.

The European Auto-Oil Programme. Report by the Directorate Generals for Industry; Energy; and Environment, Civil Protection & Nuclear Safety of the European Commission.

COM (96) 248. Auto/Oil Programme. October 1996.

Bates S et al (1996) – The Attainment of Stage III Gasoline European Emission Limits Utilising Advanced Catalyst Technology. Society of Automotive Engineers, No. 961897.

Collins N R et al (1996) – Catalyst Improvements to Meet European Stage III and ULEV Emissions Criteria. Society of Automotive Engineers, No. 960799.

Otto E, Albrecht F and Liebl J (1998) – The Development of BMW Catalyst Concepts for LEV/ULEV and EU III/IV Legislations 6 Cylinder Engine with Close Coupled Main Catalyst. Society of Automotive Engineers, No. 980418.

Hammerle R H et al. (Ford Motor Co. and FEV Motorentechnik GmbH & Co Kg (1995). Emissions from Diesel Vehicles with and without Lean NO_x and Oxidation Catalysts and Particulate Traps. Society of Automotive Engineers, No. 952391.

Havenith C, Needham J R et al (1994) – Low Emission Heavy-Duty Diesel Engine for Europe. Society of Automotive Engineers, No. 932959.

Kawanami M, Horiuchi M, Klein H, Jenkins M (1998) – Development of Oxidation and De-NO_x Catalyst for High Temperature Exhaust Diesel Trucks. Society of Automotive Engineers, No. 981196.

Touche-Ross (1995) – A Cost-Effectiveness Study of the Various Measures that are Likely to Reduce Pollutant Emissions from Road Vehicles for the year 2010. Final Report for The European Commission, DG III

Waltner A, Loose G, Hirschmann A, Mussmann L (1998) – Development of Close-Coupled Catalyst Systems for European Driving Conditions. Society of Automotive Engineers, No. 980663.

Havenith C, Needham J R et al (1994) – Low Emission Heavy Duty Diesel Engine for Europe. Society of Automotive Engineers, No. 93295.

Havenith C, Such C H, Porter B C, Nicol A J (1997) – Demonstration of a Euro 3 Heavy-Duty Diesel Engine Using Exhaust Gas Recirculation. VDI - Verlag GmbH, Issue 306/PT2, pp. 397-419.

Otto E, Albrecht F and Liebl J (1998) – The Development of BMW Catalyst Concepts for LEV / ULEV and EU III / IV Legislations 6 Cylinder Engine with Close Coupled main Catalyst. Society of Automotive Engineers, No. 980413.

Ueno H, Furutani T, Nagami T, Aono N (1998) – Development of Catalyst for Diesel Engine. Society of Automotive Engineer, No. 980195.

Dings J M W (1996) – Kosten en Milieu-Effecten van Technische Maatregelen in het Verkeer. Centrum voor Energiebesparing en Schone Technologie

Kawanami M, Horiuchi M, Klein H, Jenkins M (1998) – Development of Oxidation and de-NO_x Catalyst for High Temperature Exhaust Diesel Trucks. Society of Automotive Engineers, No. 981196.

Papameletiou D (1994) – Review and assessment of energy technologies potentially available in a long-term horizon (2020 and beyond). Fuel Cells. Final Report. IPTS.

Kalhammer F et al (July 1998) – Status and Prospects on Fuel Cells as Automobile Engines. Fuel Cell Technical Advisory Group.

Fossil Fuel Combustion in the Energy Sector

Council Directive 88/609/EEC on the Limitation of Certain Pollutants into the Air from Large Combustion Plants. OJ No L 336, 7.12.88, p. 1.

Environmental Resources Management (ERM, July 1997) – Revision of the Council Directive of 24/11/88 (88/609/EEC) on the limitation of emissions of certain pollutants into the air from large combustion plants: cost benefit analysis of this revision. Final Report for the European Commission, DG-XI.

U.S. Environmental Protection Agency – Alternative control techniques document. NO_x emissions from Industrial / Commercial / Institutional (ICI) boilers. EPA, 453/R-94-022.

U.S. Environmental Protection Agency (March 1994) – Alternative Control Techniques Document – NO_x Emissions from Utilities Boilers. EPA, Research Triangle Park, NC, USA.

European Commission – Guide to the approximation of European Union Environmental Legislation.

OECD Documents (1993) – Advanced Emission Controls for Power Plants.

Proposal for a Council Directive amending Directive 88/609/EEC on the Limitation of Emissions of Certain Pollutants into the Air from Large Combustion Plants. OJ No C 300/98, p. 6.

Amann M et al (1998) – Cost-effective control of acidification and ground-level ozone. 4th Interim Report to the European Commission. DG-XI. IIASA.

Breihofer D et al (1991) – Emission Control of SO_2, NO_x and VOC at Stationary Sources in the Federal Republic of Germany, IIP. University of Karlsruhe, Karlsruhe, Germany

Rentz O, Schleef H. J, Dorn P, Sasse H. and Ute K (1997) – Emission Control at Stationary Sources in the Federal Republic of Germany - Volume I, Sulphur Oxide Emission Control. DFIU, University of Karlsruhe, Karlsruhe, Germany.

Rentz O, Ribeiro J F (1995) – Techno-economic analysis of the SCR plant for NO_x abatement. Investigation into the optimisation potential of catalyst renewal strategies in SCR plants. Verlag Shaker, Aachen.

Soud H H, Fukasawa K (1996) – Developments in NO_x abatement and control. IEA Coal Research.

Soud H (1995) – Suppliers of FGD and NO_x Control Systems. IEA Coal Research.

Takeshita M (1995) – Air pollution costs for coal-fired power stations. IEA Coal Research.

Solvent sector

Hein J, Kippelen C, Schultmann F, Zundel T, Rentz O (1994) – Assessment of the Cost Involved with the Commission's Draft Proposal for a Directive on the Limitation of the Organic Solvent Emissions from the Industrial Sectors. Final Report for the European Commission.

4.6 References

Council Directive 1999/13/EC on the Limitation of Emissions of Volatile Organic Compounds Due to the Use of Organic Solvents in Certain Activities and Installations. OJ No. L 85/99, pg. 1.

Jourdan M and Rentz O (1991) – Reduction of the VOC from Dry-Cleaning Facilities. Report to the EC, DG XI. Karlruhe.

Macdonald E K, Marlowe I T and Woodfield M J – Control of Emissions of VOC from the Large-Scale Varnishing of Car Bodies. Final Report for the EC, DG XI. Report EUR 13568.

Heslinga D C – Solvent Emissions from Industrial and Private Use. Part 3: Metal Degreasing. Final Report for the EC, DG XI. Report EUR 13570.

den Hartog J C and Locher K (1992) – Study of the Implementation of an EC Policy for Reducing VOC Emissions from the Private and Architectural Uses of Paints and Varnishes (phase II). Final Report for DG XI.

U.S. Environmental Protection Agency (1993) – Control of Volatile Organic Compound Emission from Offset Lithographic Printing. EPA-453/D-95-001.

U.S. Environmental Protection Agency (1994) – Control of Volatile Organic Compounds from Batch Processes. Alternative Control Techniques Information Document. EPA-450/R-94-020.

U.S. Environmental Protection Agency (1994) – Alternative Control Techniques Document: Surface Coating Operations at Shipbuilding and Ship Repair Facilities. EPA-453/R-94-032.

U.S. Environmental Protection Agency (1994) – Surface Coating Operations at Shipbuilding and Ship Repair Facilities. Background Information for Proposed Standards. EPA-453/D-94-011a.

U.S. Environmental Protection Agency (1996) – Control of Volatile Organic Compounds from Wood Furniture Manufacturing Operations. EPA-453/R-96-007.

U.S. Environmental Protection Agency (1993) – Air Emissions and Control Technology for Leather Tanning and Finishing Operations. EPA-453/R-93-025.

U.S. Environmental Protection Agency (1993) – Alternative Control Technology Document. Control of VOC Emissions from the Application of Agricultural Pesticides. EPA-453/R-92-011.

Anderson K, Burnett S et al (1996) – Evaluating the Costs of Implementing the European Commission's Proposed Solvents Directive and the Scope for Using Economic Instruments. ASPINWALL/Nera.

Rentz O et al (1990) – VOC Task Force. Emissions of VOC from Stationary Sources and Possibilities of their Control. Karlruhe.

IIASA (1997) – Cost-effective control of acidification and ground-level ozone. Final report for the European Commission.

Caliandro B (1994) – Control of Technology Options and Costs for Reducing Volatile Organic Compounds. IIASA, WP-94-80.

European Commission (1994) – Feasibility Study on the Implementation of Economic Measures to Reduce Emissions of Organic Solvents. Final Report.

Chem System Ltd and ERM Economics (1996) – Costs and Benefits of the Reduction of VOC Emissions from Industry. Main Report.

Chem System Ltd and ERM Economics (1996) – Costs and Benefits of the Reduction of VOC Emissions from Industry: Annex on Cost of Control Options for Industry Sectors. Supplementary Report Part 1.

Chem System Ltd and ERM Economics (1996) – Impacts of VOC Controls on Competitiveness of Selected Industries. Supplementary Report Part 2.

Chem System Ltd and ERM Economics (1996) – Costs and Benefits of the Reduction of VOC Emissions from Industry: A Preliminary Review of the Benefits from Reduced Ground-Level Ozone Concentrations. Supplementary Report Part 3.

Klimont Z, Amann M and Cofala J (1997) – Estimating Cost for Controlling Emissions of Volatile Organic Compounds (VOC) from Stationary Sources in Europe. IIASA.

CONCAWE (1987) – Cost Effectiveness of Hydrocarbon Emission Controls in Refineries from Crude Oil Receipt to Product Dispatch. Report no. 87/52.

CONCAWE (1988) – The Control of Vehicle Evaporative and Refuelling Emissions – the "On-Board" System. Report no 88/62.
CONCAWE (1992) – VOC Running Losses from Canister Vehicles. Report no. 92/53.
CONCAWE (1990) – VOC Emissions from Gasoline Distribution and Service Stations in Western Europe. Control Technology and Cost Effectiveness. Report no. 90/52.
CONCAWE (1990) – Closing the Gasoline System – Control of Gasoline Emissions from the Distribution System and Vehicles. Report no. 3/90.
CONCAWE (1987) – Volatile Organic Compound Emissions in Western Europe: Control Options and their Cost-Effectiveness for Gasoline Vehicles, Distribution and Refining. Report no. 6/87.

Acknowledgement

The authors acknowledge the instrumental contribution of Dr. Chris Hendriks in the initial phase of the project, in particular regarding the design and the set-up of a project internal database on emission sources and emission control measures

5 Scenarios of Future Development

Stefan Reis and Rainer Friedrich

5.1 Methodology for Emission Projection

5.1.1 Approach

The prime objective for projecting the development of ozone precursor emissions for a future year was to assess the impacts of policies and legislation in place or in pipeline and to estimate the scope of the ozone problem under these conditions. The projection has to serve a number of purposes, mainly to

- provide the emission database to calculate ozone concentrations on regional and local scale in the trend year 2010,
- define a trend scenario, taking into account the legislative and technological framework and thus setting the options and limitations for further emission abatement activities,
- allow to assess efficiency of already implemented measures in terms of cost-effectiveness and ability to achieve the indicated targets, and finally
- to model the effects of structural and behavioural changes on the environmental problem under investigation.

For this study, a hybrid approach was taken, accounting for both the need of a detailed assessment and the limitations due to lacking data on implementation degrees or sectoral structures, while covering all relevant sectors and the whole of the EU15 countries. This approach is described in detail in the following sections.

5.1.2 Emission Projection – Basic Methodology

A major problem for the projection of future emission levels is that emission data is delivered to CORINAIR by all countries involved without providing the complete set of meta-data, which has been used for the calculation. Thus, it is very difficult to obtain information about e.g. fleet compositions, the age structure of power plants in the energy sector, or the amount of solvents used in a specific SNAP activity.

In general, two main approaches can be distinguished, one being based on socio-economic developments, the other technology based, but it has been proven that a combination of both approaches has to be used, if a harmonised projection of emissions from all sectors has to be done. One crucial aspect of each projection is the selection of appropriate emission factors (EF), since these process or technology specific factors relating activity rates to emissions are subject to significant changes over time. Either technological development, or legislative requirements for abatement technologies heavily influence these emission factors and thus the specific emissions per activity rate. The basic formula of emission calculation, which can be applied to the projection of future levels as well, shows this dependency:

$$E = A \bullet EF$$

with
- E = emission level (t)
- A = activity rate (activity units)
- EF = emission factor (t / activity units)

For the projection, it is vital to assess the future activity rates, such as energy demand, kilometers driven per vehicle, or amount of organic solvents used, as well as the emission factor for a future technology, which might differ considerably from that of the base year. In the CORINAIR/EMEP Emission Inventory Guidebook current emission factors are given for all sectors, but it has been necessary to use additional factors given by other information sources, e.g. the German *Emission Factor Handbook* to improve data quality by using state-of-the-art research findings.

Wherever possible, the guidelines provided by this handbook chapter have been used for the emission projection in this study, as it is described in the subsequent section.

5.2
Driving Forces of Emission Development

The approach used here for generating the projections of future emissions includes both, the assessment of future activity levels and the penetration of abatement technologies and their impact on specific emission factors. The penetration velocity of these technologies is highly dependant on the legislative framework, as can be seen in the road transport sector, where vehicle fleets show growing implementation degrees of equipment according to the increasing stringency of the EURO emission standards for road vehicles.

Future activity levels are determined by many different aspects, but for the projection, a set of indicators can be used that show sufficient correlation with the activity to be addressed. These proxies for projection are described in detail in the following sections.

5.2.1
Societal and Demographic Trends

The first proxy to be taken into account is that of the population development, since many environmental problems are directly related to urbanisation, individual or public transport demand, or energy demand. *Fig. 5.1* shows the projected growth or decrease in population in 2010 compared to 1990 for all EU Member States and as an EU15 average.

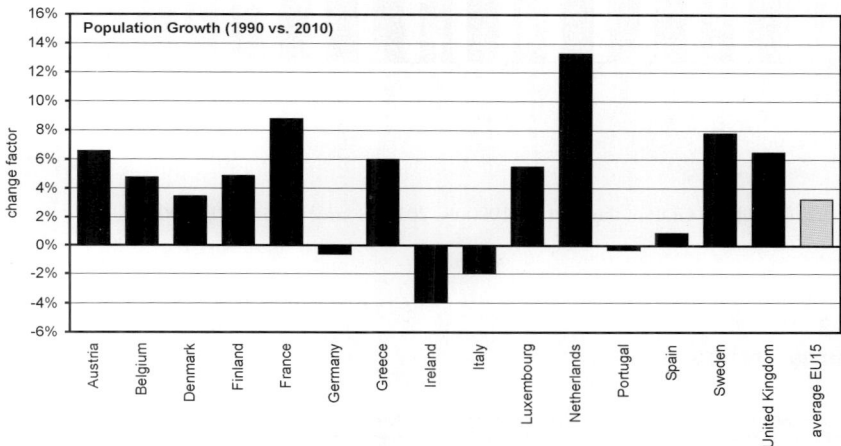

Fig. 5.1. Population growth in the EU15 – 1990 vs. 2010 (Source: EUROSTAT 1995)

This proxy was used to project activities such as the use of solvents from domestic use, or from paint application in construction and building.

5.2.2
Economic Trends

In addition to the growth of population, economic activity levels have to be taken into account, having an impact on emissions from production processes, industrial energy demand and the demand for services. The development of GDP (*Fig. 5.2*) was selected as a proxy, reflecting the different economic states in the base year and the anticipated growth rates to the trend year.

This proxy was used to project emissions from industrial energy demand, production processes and industrial applications of solvent use. As can be seen, Ireland and Portugal have comparatively high anticipated growth rates, catching up to the standards of the more industrialised EU countries.

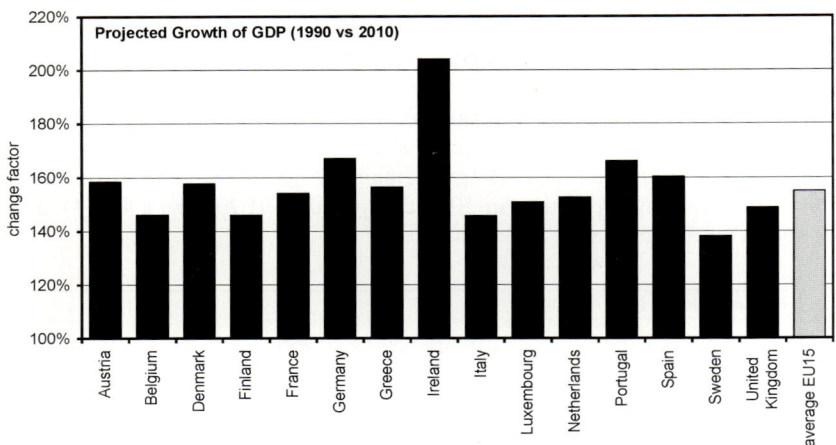

Fig. 5.2. Growth of GDP in the EU15 – 1990 vs. 2010 (Source: DG XVII 1996)

5.2.3
Energy Trends

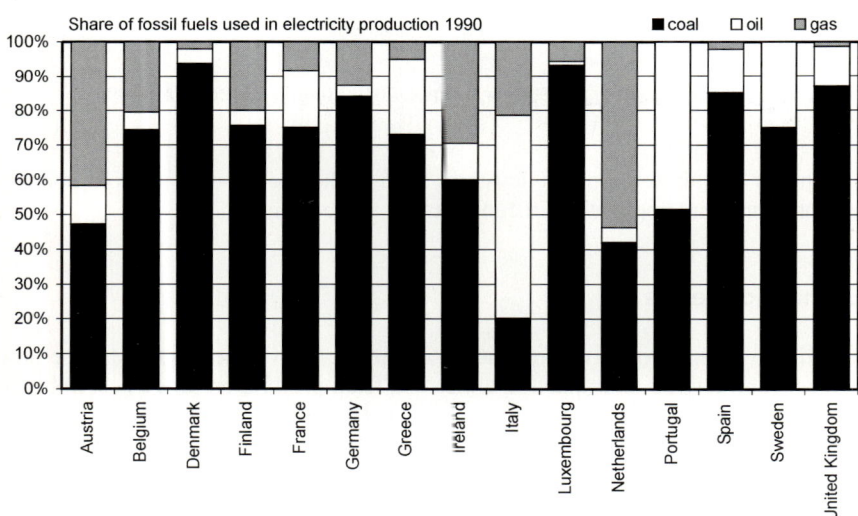

Fig. 5.3. Shares of fossil fuels used in electricity production in EU15 countries 1990 (Source: DGXVII 1996)

Since energy conversion and the use of fossil fuels in different sectors are responsible for a large share of ozone precursor emissions, it is important to thoroughly investigate trends in fuel demand and fuel shares. *Fig. 5.3* shows that the shares of fossil fuel types used in electricity production vary considerably between individual countries.

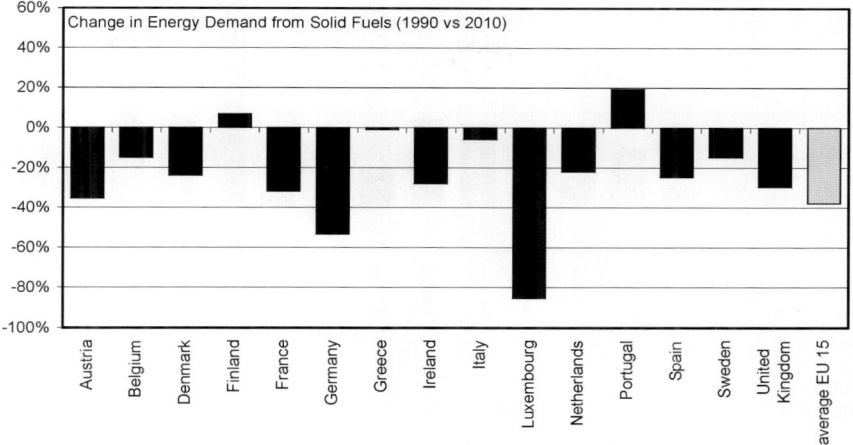

Fig. 5.4. Changes in energy demand from solid fuels – 1990 vs. 2010 (Source: DG XVII 1996)

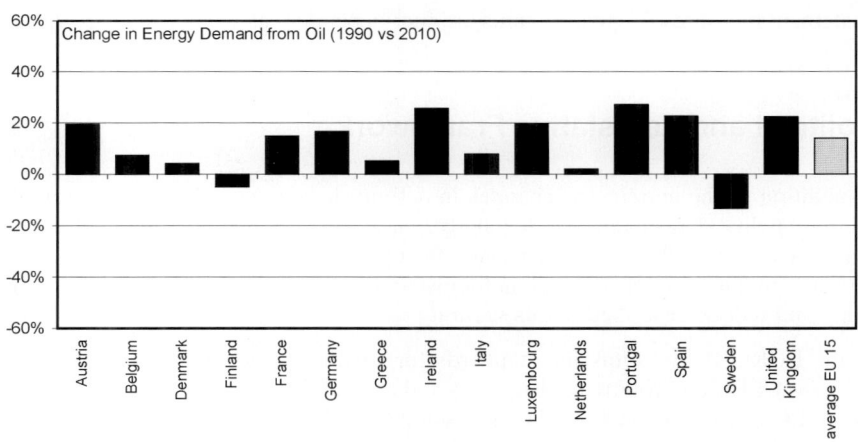

Fig. 5.5. Changes in energy demand from oil – 1990 vs. 2010 (Source: DG XVII 1996)

Figs. 5.4 to *5.6* clearly depict a trend of a decreasing consumption of solid fossil fuels (mainly hard coal and lignite) in favour of an increase of natural gas use, while oil consumption shows a slight increase. The combustion of natural gas is considered to be significantly cleaner than the use of hard coal or lignite and many new power plants being commissioned in the EU are natural gas fired.

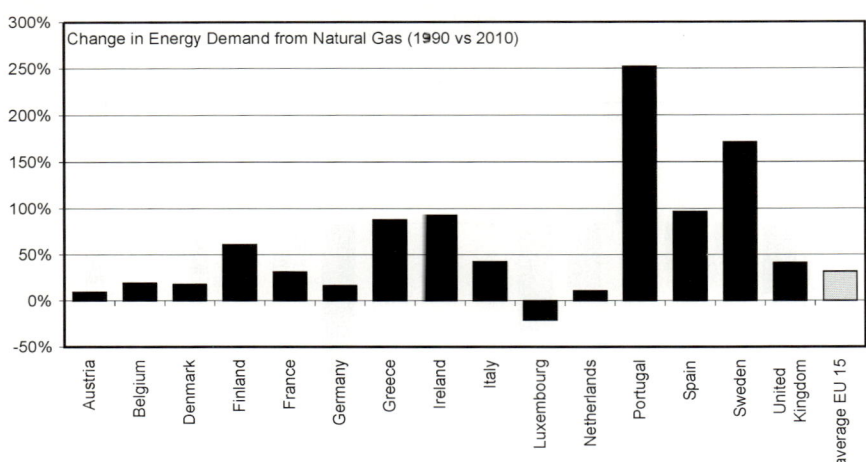

Fig. 5.6. Changes in energy demand from natural gas – 1990 vs. 2010 (Source: DG XVII 1996)

This detailed projection of the fuel mix and the development of fuel shares is important, since the NO_x emission factors of solid fuels, oil and natural gas are significantly different. Furthermore, costs and reduction efficiencies of emission abatement measures for power plants are fuel-dependent as well (*see Chap. 4.3*).

5.3
Political and Legislative Framework

In addition to the anticipated changes in activity levels described in the previous section, political decision and legislative measures can have a major effect on emissions, namely the specific emission factors of processes or plants.

Three major areas of political and legislative action have been identified in this study and will be described in this section

- the EURO I – IV emission standards for road transport vehicles (based upon 91/441/EEC and following directives and regulations, see *Sect. 4.2*)
- the EC Large Combustion Plant Directive (88/609/EEC, see *Sect. 4.3*)
- the EC Solvent Use Directive (97/C99/02, see *Sect. 4.4*)

These areas cover the most important sectors of ozone precursor emissions and have been adopted in recent years, they will develop their full impact either before the year 2010, or will at least have a significant effect until then. The individual directives and standards have to be seen in the context of the European Air Quality Framework Directive described in *Chap. 2*.

In addition to these most important legislative activities, a large number of regulations and limit values for specific substances, other sectors and groups of emitters have been taken into account as well.

5.3.1
Sectoral Air Pollution Control Policies

Road Transport

Table 5.1. Overview on legislation concerning road transport vehicles in the EU (and related ECE regulations)

EU Directive	Equiv. ECE Regulation	Vehicle Type and Emissions Control
70/156/EEC		Type approval framework Directive
70/220/EEC	ECE R 15	Exhaust emissions for gasoline passenger cars and light duty vehicles
72/306/EEC	ECE R 24.03	Heavy duty diesel black smoke emissions
74/290/EEC	ECE R 15.01	Exhaust emissions for gasoline passenger cars and light duty vehicles
77/102/EEC	ECE R 15.02	Exhaust emissions for gasoline passenger cars and light duty vehicles
77/143/EEC		In-service emissions testing
78/665/EEC	ECE R 15.03	Exhaust emissions for gasoline passenger cars and light duty vehicles
83/351/EEC	ECE R 15.04	Exhaust emissions for gasoline and diesel passenger cars and light duty vehicles
87/77/EEC	ECE R 49.01	Heavy duty diesel exhaust emissions
88/77/EEC	ECE R 49.01	Exhaust emissions from heavy duty diesels
88/436/EEC		Revised PM requirements for diesel passenger cars
88/449/EEC		In-service emissions testing
89/458/EEC		Revised CO and HC+ NO_x limits for passenger cars, implemented by 91/441/EEC
91/441/EEC	ECE R 83.01	Passenger cars; revised exhaust emissions plus evaporative emissions by ECE R15+EUDC cycles (R 83 Type Approvals B and C for gasoline and diesel respectively)
91/542/EEC		EU Clean Lorry Directive for heavy duty diesel exhaust emissions
92/55/EEC		In-service emissions testing
93/59/EEC		Exhaust emissions for light commercial vehicles (M_1 and N_1)
93/116/EC		CO_2 and fuel consumption reporting for passenger cars
94/12/EC		Passenger cars; revised exhaust emissions standards
96/1/EC		Amendments to 88/77/EEC (Production Conformity, PM for „small engines")
96/27/EC		Type approval of motor vehicles
96/69/EC		Amends 70/220 & 93/59 exhaust emissions for passenger cars and LDV

Table 5.1 presents a consistent historical overview on legislation concerning air pollutant emissions from road transport vehicles, dating back to the year 1970. For NO_x and NMVOC emissions, the current EURO emission standards are the most important regulations, at full implementation reducing up to 96 % of NO_x and 98 % of NMVOC emissions from vehicles, depending on the fuel used (*see Table 5.2*). According to the typical fleet renewal rates in each country, the EURO III and IV standards will not yet have reached full implementation in the trend scenario for 2010, with average lifetimes of a passenger car varying between less than 10 and over 15 years in individual countries. A full implementation of the EURO IV standard for all vehicle types will lead to a significant reduction of emissions from the road transport sector (about 80% of total sectoral NO_x and 88% of sectoral NMVOC emissions). Further legislation for this sector includes quality standards for transport fuel and the related activities of the distribution of fuels, such as Stage I/II of the Directive for the distribution of gasoline, covering as well service stations (94/63/EC, Stage I/II).

The amount of directives and regulations reflects the importance of this sector for air quality control, especially since transport demand is projected to grow significantly in the near future. Thus, the reductions shown in *Table 5.2* are offset to some extent by the growth in vehicle fleets, while annual mileage per vehicle is regarded to remain more or less constant. In the case of NO_x emissions from heavy duty vehicles, even an overall increase of emissions is projected for the trend scenario, reflecting a major growth in freight transport in most of the EC Member States, which outweighs the reductions achieved by the penetration of EURO IV compliant HDVs until then.

Table 5.2. Emission reductions achievable by *EURO 1 – 4* standards related to pre-EURO levels

		NO_x		
standard impl. year	EURO 1 1990	EURO 2 1996	EURO 3 2001	EURO 4 2005
PC *gasoline*	-81.2	-94.0	-94.8	-96.4
PC *diesel*	-46.3	-59.8	-57.0	-74.2
LDV *gasoline*	-82.7	-89.6	-94.8	-96.3
LDV *diesel*	-4.4	-1.9	-70.4	-73.1
HDV	-11.8	-21.6	-56.5	-76.5
		NMVOC		
PC *gasoline*	-87.6	-92.4	-94.9	-98.5
PC *diesel*	-54.8	-66.7	-78.6	-84.8
LDV *gasoline*	-87.6	-39.5	-94.7	-96.3
LDV *diesel*	-69.7	-76.2	-88.0	-97.7
HDV	-70.3	-32.0	-95.6	-92.6

Source: IPTS Analysis (see C*hap. 4)*

It has to be noted that these directives and regulations usually set an emission standard, not requiring specific equipment or technologies for compliance. The current state-of-the-art of technology options being used to achieve these standards are described in detail in *Chap. 4*.

Energy Sector

The *Large Combustion Plants (LCP) Directive* was introduced mainly to reduce the emissions of sulfur (SO_2) and nitrogen oxides from large combustion plants, which contribute by far the largest share of any individual source group to the emissions of these pollutants (about 68 % of SO_2 and 27 % of NO_x emissions, Radunsky and Ritter 1996). The Directive uses the approach of *Best Available Technology* (BAT) to install emission limits for new large combustion plants to be commissioned as well as for existing plants to comply with by a target year. The development of this Directive was related to the UNECE CLRTAP protocols (see *Chap. 2*) and to other EC Directives, e.g. regarding the sulfur content of fuels (72/116/EEC) and the *Framework Directive on Integrated Pollution Prevention and Control* (IPPC), that was adopted in 1996. In terms of environmental targets, the LCP Directive is expected to aid in accomplishing the targets of the Commissions' *Community Strategy to Combat Acidification*.

For this study, it was assumed that all countries would follow the requirements of the LCP Directive and that new plants would use BAT (mainly primary measures or SCR) to meet the limit values. For the commissioning of new plants, the projected future energy demand was used in addition to EURPROG data to assess the share of existing 1990 plants in 2010. This distinction was necessary to assess the costs for different abatement options. The high investment costs and long lifetimes (between 30 and 40 years for a typical LCP) make retrofitting feasible, other than in the transport sector, where a retrofit of cars in use is not regarded to be an option for advanced emission control technology.

An investigation carried out by ERM (1997) assumed a reduction of NO_x by 54 % relative to 1990 for the EU15, while the projection of this study amounted to a reduction of 47 %, having taken into account a higher implementation degree of emission abatement equipment in operation in Germany in the base year 1990.

5.3.2
Legislation on Specific Substances

In addition to legislation and activities concerning specific sectors as described in the previous section, the European Community directly addressed emissions from the use of organic solvents with the EC *Solvent Directive* (97/C99/02).

Aimed at setting emission ceilings and reduction targets for individual sectors of solvent use (see *Table 5.3*), the Annex to this Directive names each sector to be regulated and set specific targets to be met by a specified target year. EU Member States have to install legislation to comply with the Directive by the end of 1999. New installations are required to comply with the Directive from the start, whereas for existing installations compliance is required by October 30, 2007. EC assessment assumed a possible reduction of 57 % (relative to 1990) by 2007.

Since the projection was made for the year 2010, the Directive should have full impact on the emissions by then, affecting all installations operating in the trend year. The assessment of emission reduction potentials, however, is immensely difficult due to the scarce date available on the solvent using sector. Expertise from previous studies at IER (cf. Obermeier at al 1997, Berner et al 1996) were used along with feedback from the IPTS *Expert Survey on the Solvent Sector* to determine the potential emission reduction for all EU 15 Member States. It has been assumed that only few countries, e.g. Germany, Austria, the Netherlands, Denmark and Sweden have already started to implement emission control equipment in this sector yet, according to national legislation, thus the main activities will be taken in the coming years. Projected reduction of NMVOC emissions from the solvent use sector (SNAP 6) ranges between 18 % and 44 %, reflecting the differing shares of source activities within the EU countries, with 61 % of total emissions from this sector being covered by the *Solvent Directive*.

Table 5.3. Activities covered by the EC Solvent Directive

Activities regulated by the Directive 97/C99/02	Annex
adhesive coating	XVII
coating of	
films, paper, textiles, fabric	XIII
metallic and plastics surfaces	XI
vehicles	V, VI, VII, VIII
coils	X
leather	XVI
adhesive coating	XVII
wooden surfaces	XII
dry cleaning	XIV
impregnation of wood	XV
manufacture of coatings	XVIII
pharmaceutical processes	XXI
printing processes	III
rubber processing	XIX
surface cleaning	IV
vegetable oil extraction	XX
vehicle refinishing	IX

The Solvent Directive is implemented by the means of emission limits being set for each sector, either a combination of *process* and *fugitive emission limits* or a *total emission limit*. *Process emission limits* describe concentration limits applying to VOC emissions from contained sources (between 20 and 150 mg/m^3 depending on the sector and specific solvent consumption). *Fugitive emission limits* target uncaptured VOC emissions and are expressed as a percentage of the solvent input.

Between 5% to 45% (depending on the sector and the solvent consumption) may be emitted. Finally, *total emission limits* apply to industry sectors, setting a fixed emission limit per unit of production (e.g. for several coating sectors, dry cleaning, wood impregnation and vegetable oil extraction.

5.4 Detailed Sectoral Projection

5.4.1 Emissions from the Energy Sector

Projection of Future Activity Levels

Future energy demand was taken from the *conventional wisdom* scenario of DG XVII, using the projection of the growing input of fossil fuels into power generation (instead of growing final energy demand to take into account NO_x-neutral sources like nuclear or hydro power).

Emission Abatement Activities

For the all power plants in this sector, the requirements of *the Large Combustion Plant Directive* of the EC have been taken into account. Data for each country was harmonised similar to the approach taken by IIASA (Amann et al 1996), selecting the most stringent use of air pollution control equipment for each plant type. To comply with the LCP Directive, countries are required to equip the LCPs with primary measures (PM). In addition to that, some EU member states' regulations make the use *of Selective Catalytic Reduction*-installations (SCR) mandatory.

The projection has focussed on activities with installed capacities >300 MW, which contribute more than 97 % of total NO_x emissions in SNAP 1 (CORINAIR 90). The sectoral contribution to NMVOC emissions is very low and has not been subject to investigations, but the increasing use of natural gas in energy production will probably lead to a slight decrease in fuel related NMVOC emissions. According to the trend scenario for 2010, total EU_{15} NO_x emissions from SNAP 1 will be reduced by 47 % relative to CORINAIR 90 emissions.

5.4.2
Emissions from Residential and Commercial Combustion

Projection of Future Activity Levels

This sector is marked by a highly diffuse structure, containing emission sources from small household heating systems towards medium sized plants for commercial and institutional utilisation. Being used mainly for space heating and process heat (water etc.) purposes, no major change in activity levels per capita can be expected, while improved insulation and increasing energy efficiency in small boilers will even lead to a slight decrease in emissions.

Emission Abatement Activities

Above all, technological improvements such as increasing efficiency in boilers, improved insulation of buildings and the application of *low-NO_x*-burners and other primary measures (cf. *Chap. 4*). will have some impact on emissions in 2010. The trend scenario assumes a slight reduction of NO_x (5.1 %), but an even more significant reduction of NMVOC (9.3 %), which is primarily caused by the fuel switch from solid fuels and oil towards natural gas.

5.4.3
Emissions from Industrial Combustion

Projection of Future Activity Levels

Industrial combustion plants emit the second largest share of NO_x emissions from stationary sources. For the projection of future activity levels, the growth in energy demand within the energy sector was taken from the conventional wisdom scenario of *Energy in Europe to 2020*. While combustion plants contribute about 53 % of total sectoral NO_x emissions various industrial processes (SNAP 3.3) have been identified as a major source as well. For these processes it is assumed that an increase in emissions due to a growth in industrial production is offset by improved efficiency on a process level, keeping emissions constant on the CORINAIR 90 level. Thus, only a slight reduction is projected, reducing NO_x emissions from this sector by 16 %, mainly from the LCP Directive leading to the installation of emission control measures in industrial LPCs.

Emission Abatement Activities

For the identified combustion plants in the industry sector, the same regulations apply as for large combustion plants (LCP Directive).The implementation degree of emission control equipment has been calculated accordingly, leading to an overall sectoral emission reduction (see above). On the process level, no major emission reduction was assumed for the trend scenario. However, selected processes with significant NO_x emissions will be subject to the implementation of further reduction measures beyond the trend, e.g. cement production and iron and steel production.

5.4.4
Emissions from Industry Processes

Projection of Future Activity Levels

While NO_x emissions from Industry Processes were not relevant in 1990 (1.4% of total EU_{15} according to CORINAIR 90), the share of NMVOC emissions was considerably higher, contributing 6.7% of the EU total. For the trend scenario, no major change in this sector was projected, assuming that an increase in emissions due to a growth in industrial production was offset by higher efficiency on process level and current legislative activities to control emissions from specific industrial sources. Furthermore, data on each individual process is often restricted and thus flow rates, inputs and process details cannot be used to assess the reduction potential.

Emission Abatement Activities

For the trend scenario, no technology measures to control emissions from this sector was taken into account, but for selected processes, measures for an additional reduction of emissions have been identified.

5.4.5
Emissions from Fuel Handling

Projection of Future Activity Levels

The fuel handling sector is a major contributor to NMVOC emissions (7.3 % of total EU_{15} emissions according to CORINAIR 90). Within this sector, the handling of liquid fuels was identified to be the most significant source, concentrating on the distribution path of gasoline from refineries to service stations. As a proxy to assess the future development of this source, the growth in demand for fossil fuels was used (Energy in Europe to 2020, DG XVII). Specific growth factors have been identified for solid fuels, oil, natural gas and gasoline (see *Fig. 5.7*).

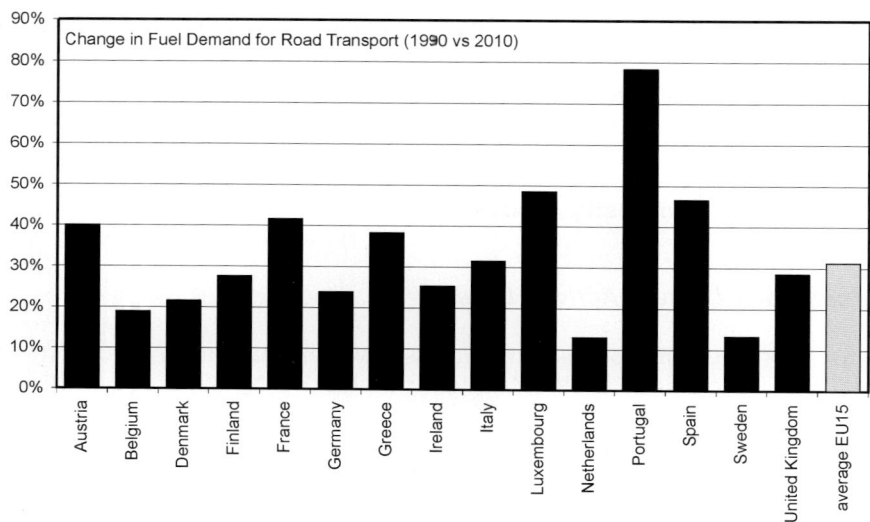

Fig. 5.7. Changes in fuel demand for the transport sector in the EU15 – 1990 vs. 2010 (Source: EC DG XVII)

Emission Abatement Activities

Following directive 94/63/EC, activities to reduce NMVOC emissions have been taken into account at the refinery and depot level (Stage IA) and for service stations (Stage IB). Furthermore, Stage II controls have been assumed to be implemented for the largest share of service stations, resulting in an overall reduction of 45 % within gasoline distribution (57 % for large service stations, 15 % – 1 % for medium and small service stations with/without derogation). Due to the projected increase in fuel demand, total sectoral emissions of NMVOC only show a reduction of 26.3 % compared to CORINAIR 90.

5.4.6
Emissions from Solvent Use

Projection of Future Activity Levels

With a share of 28.7 % of total EU_{15} NMVOC emissions, solvent use was identified as the second largest contributor. And since the bulk of activities covered by this sector is related to industrial activity, growth assumptions for the countries' Gross Domestic Product (GDP) have been used as proxies to project future activity levels. For the *domestic use* of solvents or paints, population development was used instead.

Emission Abatement Activities

The EC Directive on VOC emissions from solvent use (97/C99/02) and the UN/ECE VOC protocol to the Convention on Long Range Transboundary Air Pollution (CLRTAP) have been taken into account to reflect current reduction plans in this sector. For all countries of the EU_{15}, significant reductions of emissions from solvent use have been assumed, leading to an overall decrease of NMVOC emissions from SNAP 6 by 31 % compared with CORINAIR 90. For all countries, compliance with the VOC protocol is foreseen. Although for Greece, Portugal and Ireland the relatively high growth rate of industrial production (reflected through GDP) will prove to offset abatement activities to a large extent.

5.4.7
Emissions from Road Transport

Projection of Future Activity Levels

For emissions from Road Transport, several variable have been investigated to assess the development of emissions. On the one hand, emission factors determine the amount of a pollutant emitted related to a specific activity (e.g. g/km driven), on the other hand, the size and technological composition of the vehicle fleet and a change in activity rates and patterns have are of interest. Data on size and composition of EU_{15} vehicle fleets was obtained from *Deliverable 16* of the MEET project (MEET 1997b). To be used for the generation of the trend scenario for the year 2010, this data had to be adapted to meet the sectoral aggregation level of CORINAIR 90 and to improve transparency. Thus, vehicles have been attributed to technology groups such as *pre-EURO, EURO I, EURO II, EURO III* and *EURO IV* for Passenger Cars (PCs), Light Duty Vehicles (LDVs) and Heavy Duty Vehicles (HDVs). For Two-Wheelers (TWs), 2-stroke and 4-stroke engines were distinguished into *Stage I* and *Stage II* controls for Mopeds (< 50 ccm) and *uncontrolled* and *controlled* (> 50 ccm).

Given activity rates have also been adapted to comply with this aggregated data, providing information on *km driven per vehicle and year*. The change from 1990 to 2010 levels was calculated using only modifications in emission factors and fleet composition (vehicle types and technology levels); annual mileage per vehicle was kept constant, assuming that a single vehicle would not be operated in a way much different from 1990, regarding the shares of urban, rural and highway driving patterns as well.

For PCs and LDVs a trend exists towards an increasing share of diesel operated vehicles, while for HDVs the share of gasoline fuelled vehicles was assumed to be zero. Liquified Petrol Gas (LPG) does play a minor role, being only used to some extent in Belgium, The Netherlands and Italy and was thus not being taken into account.

Projected emission levels show considerable reductions in this sector, amounting to a decrease by 39 % for NO_x and even 67.5 % for NMVOC.

Emission Abatement Activities

Together with data on technical emission abatement measures (see *Chap. 4*), IPTS provided mean emission factors for NO_x and NMVOC for each vehicle type and EURO emission standard. The emission factors used for the projection reflect the development of European legislation on emission standards for road vehicles. *Table 5.1* presents all regulations taken into account. The implementation degree of each standard was provided by MEET (Samaras et. al 1997a, b, c and 1998), assuming the timely coming into force of EURO III and EURO IV standards. Since a normal turnover of vehicle fleets was projected, there will still be only relatively few vehicles complying to EURO IV in the 2010 trend case. And since a retrofitting of vehicles with –mainly built-in– emission abatement technologies is not feasible, only activities to promote a faster turnover of the vehicle fleet can lead to a further significant reduction of emissions from road transport beyond the trend.

5.4.8
Emissions from Other Mobile Sources

Projection of Future Activity Levels

This sector proves to be rather difficult to handle, because it comprises a collection of sources, which do not have much in common. For emissions from Airports (SNAP 8.5), a significant increase can be assumed due to growing demand for air transport. However, contributing only 4.7% of total sectoral NO_x emissions, this has not been taken into account for the trend scenario. The main source group within this sector is that of off-road vehicles and machines (SNAP 8.1) with a share of 50% of total sectoral NO_x and even 60% of total sectoral NMVOC emissions. Activity levels for this source group have been assumed to remain constant, too.

Emission Abatement Activities

While other mobile sources emit a significant share of total EU_{15} NO_x emissions (12.4 %), their contribution to total EU_{15} NMVOC emissions is considerably lower (3.8 %). For the trend scenario, no major emission reduction activities have been taken into account, assuming that technological improvements especially in the field of off-road vehicles and machines will be offset by increasing activities. But, for many off-road vehicles, abatement options are similar to those of road vehicles. Thus, these sources will be taken into account in the implementation of additional abatement technologies beyond the trend case.

5.4.9
Emissions from Waste Handling

Projection of Future Activity Levels

Waste handling does only contribute marginally to total NO_x and NMVOC emissions. For the projection of trend emissions, population development was taken as a proxy variable.

Emission abatement activities

For the trend scenario for 2010, no emission abatement measures haven been taken into consideration, but especially NO_x emissions waste incineration (SNAP 9.2.0) can be significantly reduced by additional measures, as an option for further abatement activities.

5.4.10
Emissions from Agriculture (SNAP 10)

Projection of Future Activity Levels

For agriculture, the major part of NMVOC emissions are cultures with fertilizers (29 %) and excretions from animal breeding (61 %). According to IER research, activity levels in this area can be assumed to remain constant in the EU, with no major changes in the intensity of farming or animal breeding to be anticipated

Emission Abatement activities

For the trend scenario, no emission abatement has been taken into consideration.

5.5
Trend Scenario Emissions

Compiling the detailed sectoral projections, the trend scenario for 2010 shows an overall reduction of NO_x by 30 %, for NMVOC the reduction is 36 % (*fig. 5.8*), with respect to growing economies and e.g. increases in road transport a considerable decrease in emissions. Though to – achieve a significant reduction of tropospheric ozone concentrations, it will not be sufficient, thus making it necessary to identify additional measures and strategies for reduction.

As *Figs.* 5.9 and 5.10 (see *Tables* 5.4 and 5.5 as well) indicate, the relative reduction of emissions is far from homogeneous among EU15 countries, changes ranging from +1.7 to –42.4 % for NO_x and –13.1 % to 48.2 % for NMVOC. But it can be seen that the major emitters (Germany, France, Italy and the United

5 Scenarios of Future Development

Table 5.5. Anthropogenic NMVOC emissions in the trend scenario for 2010 compared with CORINAIR emission inventory data

NMVOC EU15	CORINAIR 90	CORINAIR 94[*]	Trend 2010
Austria	418.9	314.7	299.5
Belgium	364.9	342.0	194.4
Denmark	169.1	156.2	87.5
Finland	165.3	177.1	118.0
France	2.403.7	2 375.4	1 469.1
Germany	2.936.6	2 203.2	1 582.1
Greece	324.9	277.7	282.3
Ireland	180.4	105.4	132.7
Italy	2.395.8	2 320.5	1 671.3
Luxembourg	18.7	17.9	10.9
Netherlands	456.7	378.8	329.9
Portugal	205.8	285.7	156.3
Spain	1.118.8	1 247.5	882.3
Sweden	451.3	342.3	324.0
United Kingdom	2.602.0	2 362.7	1 614.6
Total EU15	**14.213.2**	**12 907.2**	**9 154.9**
change relative to CORINAIR 90		*- 9 %*	*- 36 %*

[*] in CORINAIR 94, the methodology for calculating biogenic emissions from forests (SNAP 11) and agriculture (SNAP 10) has been changed; these figures have been corrected to be comparable to CORINAIR 90 and the trend scenario for 2010

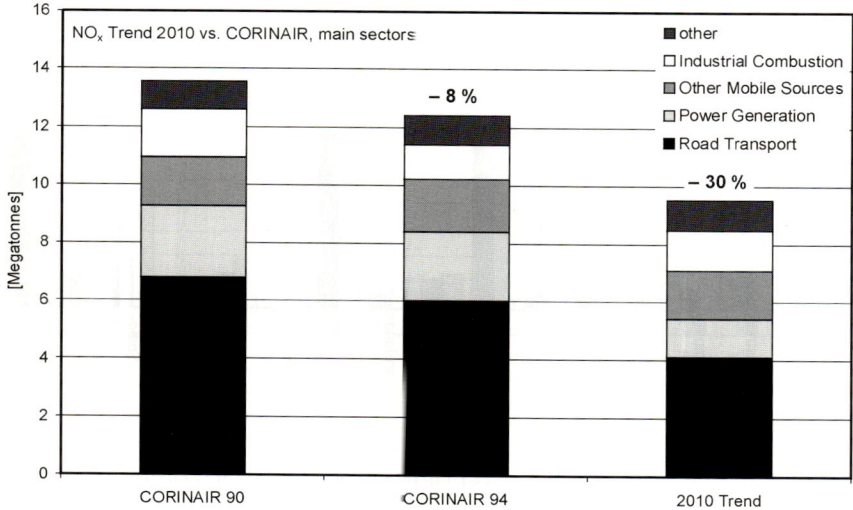

Fig. 5.11. 2010 NO_x trend emissions vs. CORINAIR emissions, main sectors

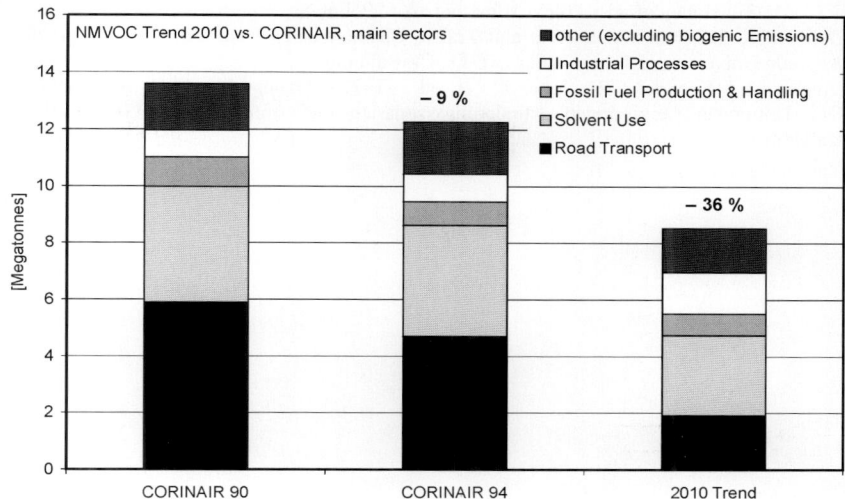

Fig. 5.12. 2010 NMVOC trend emissions vs. CORINAIR emissions, main sectors

5.6 References

Amann et al. (1996) Cost-effective Control of Acidification and Ground-Level Ozone. First Interim Report to the European Commission, DG-XI, Laxenburg
Berner P, Obermeier A, Friedrich R et al. (1996) Strategien zur Minderung der VOC-Emissionen ausgewaehlter Emittentengruppen in Baden-Wuerttemberg. FSKA-PEF 147, Karlsruhe
ERM (1996a) Revision of the Council Directive of 24/11/88 (88/609/EEC) on the Limitation of Emissions of Certain Pollutants into the Air from Large Combustion Plants: Cost Benefit Analysis of this Revision. Report to the European Commission, DGXI, London
ERM (1996b) Costs and Benefits of the Reduction from VOC Emissions from Industry. Report to the Department of Trade and Industry, London
EUROSTAT (1995) Europe's Environment – Statistical Compendium for the Dob•íš Assessment. Office for Official Publications of the European Communities, Luxembourg
DGXVII (1996) Energy in Europe – report submitted by DG XVII, Brussels
Radunsky K, Ritter M (1996) CORINAIR 1990 Summary Report 3 – Large Point Sources, ETCAE/EEA, Copenhagen
MEET (1997a) Methodologies for Estimating Air Pollutant Emissions From Transport – Deliverable 7: Average Hot Emission Factors for Passenger Cars and Light Duty Trucks, LAT-Report 9716, Aristotle University of Thessaloniki (AUT), Thessaloniki
MEET (1997b) Methodologies for Estimating Air Pollutant Emissions From Transport – Deliverable 16: Road Traffic Composition, LAT-Report 9718, Aristotle University of Thessaloniki (AUT), Thessaloniki
MEET (1997c) Methodologies for Estimating Air Pollutant Emissions From Transport – Deliverable 20: Fuel and Energy Production Emission Factors, ETSU Report No. R112, Harwell

MEET (1998) Methodologies for Estimating Air Pollutant Emissions From Transport – Deliverable 21: Emission Factors and Traffic Characteristics Data Set, LAT-Report 9802, Aristotle University of Thessaloniki (AUT), Thessaloniki

Obermeier A, Friedrich R, John C et al. (1997) Ozonproblematik im suedlichen Oberrheingraben: Emissionen, Minderungsszenarien und Immissionen. FSKA-PEF 162, Karlsruhe

6 Regional Modelling of Tropospheric Ozone

David Simpson and Jan Eiof Jonson

6.1
Introduction

Atmospheric transport and chemistry models are an essential part of assessing ozone abatement. Such models allow a quantification of both the levels of ozone concentration across the whole of Europe, and of the relative effectiveness of different control measures. In this study, two photochemical models of the European Monitoring and Evaluation Programme (EMEP[1]) have been used to investigate the effects of emission control scenarios and to assess the contribution of specific source sectors to ozone formation over Europe. In a first step, the status quo of ozone concentrations was calculated for the base year 1990 and the trend scenario 2010, showing that even though precursor emissions were reduced significantly, ozone thresholds were exceeded to a large extent. Data from these models were also used as inputs for the emission-optimisation (*Chap. 7*), economic evaluation (*Chap.s 9 & 10*), and for more local modelling (*Chap. 8*). An important application of the Lagrangian model was the provision of source-receptor-matrices for the optimisation by the means of calculating matrices to relate emissions to ozone formation for a 5-year average (see *Sect. 6.5*).

6.2
The Models

The two EMEP models used are the Lagrangian photo-oxidant model (Simpson, 1992a, 1993, 1995), and the 3-D Eulerian model (Jonson et al., 1997, 1998, 1999). A summary of the two models main characteristics is given in *Table 6.1*; more details can be found in Simpson and Jonson (1998). The Eulerian model is superior in terms of grid resolution and physical formulation, but is a relatively recent model and requires significant computer time. The Lagrangian model has formed the basis of EMEP ozone work for several years and it is much faster to

---

[1] More formally, the Cooperative Programme for Monitoring and Evaluation of the Long Range Transmission of Air Pollutants in Europe (http://www.emep.int).

run for different simulations. Further, results from this model have so far formed the basis for UNECE and EU work on the so-called NO$_x$ Protocol, the EU Auto/Oil study, as well as an Acidification and Ozone strategy (Simpson and Andersson-Sköld, 1997, Amann et al., 1999). Both models have been extensively compared with measurements (Simpson, 1992a, 1993, Simpson et al., 1998, Solberg et al., 1995, Jonson et al., 1993). A side-by-side comparison of the two EMEP models was given in Simpson and Jonson (1998). Both models were found to be roughly comparable in performance, although the 3-D model seemed to show improved performance in reproducing day-to-day ozone variations. Comparisons of the Lagrangian model with other models have been presented in Hass et al. (1997), Kuhn et al. (1998) and Andersson-Sköld and Simpson (1999).

Although *Table 6.1* points to some areas of difference between the two models, the main difference is in their basic physical structure, with the Eulerian model featuring 20 vertical levels and 50×50 km^2 horizontal resolution, whereas the Lagrangian model has just one layer and 150×150 km^2 horizontal resolution.

6.3
Emission Input to Models

The anthropogenic emissions used here are identical for the two models. Emissions are specified on the 50 km EMEP grid for each of the 11 SNAP-1 sectors of the EMEP/CORINAIR system. The emissions used have been described in *Chap. 3*, with base-case 1990, 2010 ("trend") and 2010 scenario data-sets for the EU countries. For non-EU countries emissions have been taken from EMEP data.

In the Lagrangian model a daily emission variation is applied, derived from the GENEMIS project (Friedrich, 1993). The Eulerian model uses monthly and daily (sunday-saturday) factors derived from a more recent calculation under GENEMIS (U. Schwarz, Univ. Stuttgart, *pers. comm.*).

The biogenic emission inputs are calculated internally in the models using the same land-use and E-94 methodology of Simpson et al. (1995).This should ensure similar emissions in the two models, although some differences may be created by the temperature fields which have higher spatial resolution in the Eulerian model.

VOC speciation data for the oxidant models were derived from an extensive set of data compiled in the frame of the German Tropospheric Research Project. This data set classified VOC into the 38 classes recommended by Middleton et al. (1990), including data for 34 classes of vehicle/driving modes. These VOC were grouped into more aggregated classes suitable for the oxidant models (see Andersson-Sköld and Simpson, 1999, Reis et al., 1999). For each scenario considered the VOC speciation was calculated for each SNAP category using data on fleet composition and driving modes for each country.

Table 6.1. EMEP MSC-W Lagrangian and Eulerian models, brief summary.

	Lagrangian	Eulerian	Comment
Resolution			
Horizontal	150 km	50 km	
Vertical layers	1	20 lowest layer ca. 40m	
Vertical extent	to mixing height, ca. 1-2.5 km	To 100 hPa, ca 15 km	
Emission inputs			
Source	CORINAIR/EMEP, 50 km grid		
Source-sector split	yes: 11 SNAP sectors	yes: 11 SNAP-sectors	
Temporal resolution	Daily	Monthly + daily factors	GENEMIS, see text
Biogenic Emissions	Isoprene, calculated every hour in model, E-94 methodology		Simpson et al., 1995
Chemistry			
Type	Lumped molecule. Originally Eliassen et al., 1982, but updated independently.		
Reactions	140	88	
Species	66	50	
Photolysis	Jenkin et al (1997), Derwent et al (1996)	DeMore et al, 1997, Kylling 1995	Eulerian model J-values more up-to-date
Other			
Deposition module	Resistance-based	Drag coefficient	
Met. input:	DNMI; LAM50, averaged to 150km	DNMI; LAM50	
Boundary conditions	Climatalogical, from measurements	Monthly, from global model	
CPU-requirements	ca. 1000 s	ca. 70 000 s	6-month run, CRAY T3E, 16 processors, time per processor

6.4
Statistics Used

Any run of the EMEP models results in a large amount of output data, with hourly or 6-hourly concentrations of ozone for each grid point of the model over a 6-monthly period. In order to summarise these model results over long time periods calculated changes in two statistics, AOT40 and AOT60 are presented. The AOT statistics have been defined in *Chap. 2*. These statistics have important policy implications as indicators of forest damage (AOT40, Fuhrer et al., 1997) and health (UN-ECE, 1997), and were recently used in the EU acidification and ozone strategy work (Amann et al., 1999).

It should be noted though that ACT values are very sensitive to systematic biases in both modelled and observed data (especially when ozone concentrations lie just below or above the threshold limits). Simpson et al. (1998) showed that a 10% uncertainty in measured hourly ozone can lead to differences of typically 100% and 200% (for one site 1000%) in AOT40 and AOT60 respectively.

AOT values provide a good summary of the ozone situation in a specific grid square, but it still means having many data-values: one prediction of AOT40 and AOT60 for each grid square of Europe. In order to summarise these results across all of the EU, percentiles of these AOT values have been calculated for the EU area. Thus the 100th percentile for any given scenario represents the maximum AOT value found by the model within the EU. A 90th percentile of X ppm.h means that 90% of the EU area has AOT values of less than X ppm.h. Additionally, mean AOT values have been calculated as the simple arithmetic mean of all the AOT values found over the EU area.

Some of the tables present differences in AOT40 values for given percentiles (e.g. ΔAOT40). These differences represent the difference between the given percentile value AOT values found in the emission scenario and the same percentile found in the base-case. Such ΔAOT values thus represent changes in the statistical distribution of AOT values, and do not refer to any specific grid square (e.g. the grid square with max AOT, the 100th percentile, may well be different in the base and scenario cases).

6.5
Model Calculations: 1990 Base and 2010 Trend Case

The following sections present and discuss the main model results obtained in the this study. Most of the studies with the Lagrangian model have involved simulations over 5 different summer periods, but keeping the emissions fixed at the desired scenario level. The 5 years chosen were 1989, 90, 92, 93, 94, consistent with those used in Simpson et al. (1997) and Amann et al. (1999). This procedure gives a 5-year average ozone prediction, which is superior to just using the meteorology of one year. Meteorological data for the 3-D model were only

6.5 Model Calculations: 1990 Base and 2010 Trend Case

available for only one year, 1996, so the Lagrangian model has also been run for this year, and this enables the 3-D results to be placed in a longer-term perspective.

The initial studies (base-case 1990, 2010 calculations) are similar in concept to previous investigations within the EMEP project, and further discussion of the general pattern of ozone and AOT40 across Europe, and year-to-year variations, can be found in e.g. Simpson, 1993, 1995, Simpson et al., 1998. The main new feature within this study has been the extensive use of the Eulerian model and the sector-specific emission scenarios presented in *Sect. 6.6*, which have been conducted in addition to the model evaluation of optimisation results to assess the contribution of individual sectors to ozone formation over Europe.

6.5.1
Base-Case 1990 and Trend 2010 Scenario, 5 yr Average

Plate 1 illustrates predicted AOT40c (AOT40-crops) levels for 1990 level emissions and the emissions calculated for the 2010 trend scenario. These calculations are intended to give an average picture of the effects of the control measures, and have been performed using the Lagrangian model and 5 years of meteorological data as discussed above. The model results show a significant reduction in AOT40c levels by the year 2010, but also that AOT40 levels remain well above the suggested critical threshold of 3 ppm.h in most parts of Europe.

Plate 2 shows the corresponding results for AOT40f (AOT40-forests), which uses the April-September meteorological data from the 5 summers. As for AOT40c, significant reductions are achieved by 2010 according to these calculations, but again most parts of Europe lie above the critical threshold of 10 ppm.h for forests.

In *Plate 3*, AOT60 levels are seen to fall much more dramatically between 1990 and 2010,with about 50% reductions in many areas. This more dramatic reduction reflects the greater sensitivity of AOT60 compared to AOT40, but the percentage change should be viewed with some caution as noted in *Sect. 6.4*. (it is expected that AOT60 should be more sensitive to emission control measures than AOT40, but the absolute values of AOT60 are very uncertain and subject to strong bias, so the relative effects of control measures may be significantly under-or-over-estimated).

6.5.2
Comparison of Models: Trend 2010 Scenario

Both the Lagrangian and Eulerian models have been run for the year 2010 scenario using meteorological data from April-September 1996. The calculated AOT $40_{forests}$ (*Plate 4*) is generally similar from the two models, with levels of less than 5 ppm.h in Northern and Eastern Europe, increasing to 15-25 ppm.h especially in parts of Italy. The main differences are that the 3-D model has somewhat lower AOT $40_{forests}$ over North-Western Europe, especially in Germany and the UK. This latter difference is probably due in large-part to the vertical resolution of the 3-D model. With a lower-layer thickness of ca. 40m, surface NO_x sources are much better resolved in the 3-D model than in the 1-layer Lagrangian model. This gives greater NO_x concentrations in the lowest layer and hence lower ozone, through the titration reaction $NO + O_3 \Rightarrow NO_2$.

The model comparison for year 2010 emissions is given in *Plate 5*. In view of the uncertainties in predicting AOT60 the two results are surprisingly similar: both models predict similar gradients from Northern/Eastern Europe into Central and Southern Europe, with maxima of 1.5 - 3 ppm.h. The Lagrangian model shows lower AOT60 over much of the UK and parts of Spain and Portugal, but higher over parts of Germany. As AOT60 is driven more by episodic high ozone events, often with good vertical mixing, it is not straightforward to interpret these differences in terms of vertical resolution as can be done with AOT40. The differences are likely to be due to a number of factors involving grid size, non-linearities in the chemistry, and different model setups (photolysis rates, boundary conditions, etc.)

Comparing the Lagrangian models predictions using 1996 meteorology (*Plate 4*) with the 5-year average results (*Plate 2*), it can be seen that 1996 is a year with relatively low $AOT40f$ levels. On the other hand the AOT60 levels predicted using 1996 meteorology (*Plate 5*) are in many areas higher than for the 5-year average (*Plate 3*). These results confirm both the importance of year-to-year variations, and the different characteristics of the long-term AOT40 and more episode-based AOT60. It is clear that in future, long-term 3-D modelling will be required to evaluate emission control measures in a statistically sound sense.

6.6
Model Calculations: Sector-by-Sector Emission Scenarios

In the following two sections the relative importance of the different emission sectors for ozone ($AOT40_f$ and AOT60) across Europe is investigated. All of the calculations are conducted for the year 2010, using 1996 meteorological data. In *Sect. 6.6.1*, the Eulerian model is used to look at each of the SNAP sectors 1-10 in turn. In each scenario the emissions of that sector are completely removed for all EU countries, the model re-run, and the model results obtained compared with the base-2010 results. *Sect. 6.6.2* presents a similar analysis, but now looking in more detail at the road traffic sector – with SNAP levels 7.1 (passenger car exhaust) to 7.6 (evaporative emissions), investigated with both the Eulerian and Lagrangian model.

These analyses do not attempt to say how much control is technically possible within each emissions sector, but rather compare the relative contribution of the sector as a whole to European ozone levels.

6.6.1
Emission Control Measures, SNAP sectors 1-10

Table 6.2 summarises the results of the sector-by-sector emission control runs, giving the average European-wide change in AOT40 or AOT60 associated with each emission sector. The road transport, solvent and production processes sectors are seen to account for the biggest reductions in both AOT40 and AOT60. Sectors with mainly NOx emissions (SNAP 1, 3 and 8) do contribute to AOT reductions also, but to a lesser extent than those sectors with mainly VOC emissions (SNAP 4,6) or mixed NO_x and VOC emissions (SNAP 7).

Table 6.2. Emission projection (year 2010 base-case) and mean reductions in European AOT40 (forests) and AOT60 obtained when EU sector-emissions are set to zero.

SNAP code	Emissions sector	Emissions		Mean AOT reduction [a]	
		NO_x (ktonnes)	NMVOC (ktonnes)	ΔAOT40$_f$ (ppb.h)	ΔAOT60 (ppb.h)
1	Public power, co-generation and district heating	1041	41	134	9
2	Commercial, institutional and residential heating	499	730	148	18
3	Industrial combustion	1315	79	68	4
4	Production processes	340	1450	681	72
5	Extraction and distribution of fossil fuels	110	946	357	38
6	Solvent use	0.9	2756	1000	110
7	Road transport	3990	1820	1138	100
8	Other mobile sources and machinery	1518	510	312	28
9	Waste treatment and disposal	121	244	83	14
10	Agriculture	33	0 [b]	16	1

[a] the mean ΔAOT values are calculated over all of Europe. For comparison the European mean base-case 2010 AOT40f is ca.7200 ppb.h, and mean AOT60 is 230 ppb.h.
[b] agricultural VOC emissions are set to zero for all countries, to avoid double counting forest VOC emissions from managed forest (often reported as agricultural VOC) with the biogenic emissions calculated within the model.

The spatial distribution of the AOT40$_f$ changes is illustrated in *Plates 6-7*. These clearly show that reduction of the NOx-dominated sectors results in AOT40 increases in many areas, but reduction of the VOC sectors gives only AOT40 reductions. Sectors such as road-transport, other mobile sources and waste treatment show AOT40 reductions in most areas, but increases in areas with high NO_x emissions (southern UK, BENELUX areas, north-west Italy).

The contrasting behaviour of NO_x and VOC emissions sources is a well known feature of non-linear ozone chemistry, and has been discussed further for example in Sillman et al.,1990, Pleim and Ching, 1993, or Simpson, 1995.

Unfortunately it is very difficult to model this behaviour correctly; even a reduction in grid-size is not guaranteed to give behaviour that is more realistic, just different (Pleim & Ching, 1993). However, in those affected grid squares, there is a clear need for more local-scale modelling (e.g. Moussiopoulos et al., 1999). The increased effectiveness of evaporative emissions (which are a pure VOC source) suggested by the Eulerian model is also probably a grid-size phenomenon. In high-NO_x areas VOC emissions produce more ozone, and the 3-D model has more grid cells with higher NO_x than the Lagrangian model.

6.7
Country-to-Country Calculations (Source-Receptor Relationships)

Source-receptor relationships give the change in ozone in each receptor country (or grid square) resulting from a change in emissions of either NO_x or VOC from each emitter country. These relationships are derived by running the EMEP Lagrangian model many times, each time reducing the emissions of one emitter country. The results of each run are compared to that of a base-case run (in this case the year 2010 trend scenario), in order to calculate the change (Δozone) attributable to each country. Of course, ozone production is a non-linear process, so source-receptor matrices calculated under one set of conditions are not strictly applicable in other conditions. However, several studies have demonstrated that such matrices are approximately correct over a wide range of conditions (Simpson and Malik 1996, Simpson 1992b).

Source receptor matrices have been calculated by running the EMEP model with the emissions of each EU country reduced in turn, both for a 40 % NO_x reduction and for a 40% VOC reduction. All changes are in anthropogenic emissions only, and expressed relative to the 2010 trend case. Simulations were conducted with meteorology for 5 years and averaged. The detailed model results are used as the basis of the OMEGA model (Chap. 7).

A brief summary of these results in a highly aggregated form – giving the mean AOT changes over each "receiver" country resulting from the 40% emission change in each "source" country – is shown in *tabs. 6.6 - 6.9*. The contribution of each country's NO_x and VOC emissions to European-average AOT40 and AOT60 is further illustrated in *Plate 9*. As in previous studies (Simpson and Malik 1996, Simpson et al. 1997), these country-to-country matrices reflect two main factors. Larger countries have larger emissions and in general this leads to a bigger contribution to European ozone levels. The important exception arises for countries with dense NO_x emissions in some areas (e.g. Germany, Netherlands and the UK) where the non-linear ozone titration reaction leads to negative ozone contributions. Reduction of NO_x emissions from such countries should preferably be accompanied by simultaneous VOC reductions (*see Chap. 7*).

6.8
Summary and Conclusions

In this study, two photochemical models of the European Monitoring and Evaluation Programme (EMEP) have been used to investigate the effects of various emission control options applied within the European Union. Data from these models have been used as inputs for the emission-optimisation (*Chap. 7*), economic evaluation (*Chap.s 9-10*), and for more local modelling (*Chap. 8*). In addition to that, the models were used to evaluate the results of the optimisation to ensure that concentrations calculated by the reduced form iteration model showed the correct picture.

The two models suggest that despite the improved air quality expected in 2010 as a result of currently planned measures, levels of AOT40 will still lie significantly over the critical thresholds for vegetation damage. AOT60 levels are also predicted to be above zero in 2010 in most parts of Europe, suggesting continued risk of health effects. The models further suggest that there is still substantial scope for reducing AOT40 and AOT60 levels further.

Calculations have been performed to look at each emission sector in turn, especially for the road transport sector, to evaluate their contribution to AOT40 and AOT60 over Europe. The sectors contributing most to AOT levels in this analysis were found to be road transport, solvents and industrial processes. Within the road transport sector, heavy-duty vehicles and evaporative emissions are predicted to make the largest contributions, followed by passenger-car exhaust.

Some caveats should be expressed in considering these results. The uncertainties of these calculations are quite large. One source of uncertainty is clearly the emission inventory and forecast to the year 2010. Another source lies in the model performance: as has been illustrated, the relative effectiveness of different measures do depend to some extent on the model used, reflecting at least partly the effects of grid resolution on non-linear ozone formation. These issues need to be explored further, ideally with even higher resolution models.

However, models are essential to evaluating the control measures precisely because of these non-linear interactions (ozone formation is not simply proportional to emissions). Despite their uncertainties they provide an important tool for predicting the likely effects of future control strategies. As part of this work, it has been shown that two models of very different resolution produce similar results over most of their range of prediction, especially in terms of ranking possible control measures. Where differences occur the reasons are fairly well understood. The results presented here should therefore add valuable information to the discussions concerning emission abatement.

Table 6.6. Country-to-country matrix for 40% NO$_x$ Reduction, AOT40-crops

		Source-country												
		AT	BE	DK	FI	FR	DE	GR	IE	IT	LU	NL	PO	ES	SE	UK
Receptor-country	AT	273	-15	3	6	113	102	1	1	225	0	-39	0	8	12	-32
	BE	4	-278	3	3	405	-966	0	8	6	9	-614	1	32	12	-494
	DK	1	-18	158	21	12	87	0	3	0	0	-54	0	1	108	-30
	FI	0	-1	11	66	0	5	0	0	0	0	0	0	0	60	1
	FR	5	-38	2	1	1275	-141	0	8	93	2	-118	6	167	4	-127
	DE	39	-63	15	9	183	-153	0	3	23	1	-206	0	11	26	-128
	GR	4	0	0	1	12	8	530	0	174	0	0	0	6	2	0
	IE	0	-12	2	0	24	-58	0	98	0	0	-35	0	5	3	-156
	IT	25	0	0	2	270	2	16	1	1004	0	-5	2	52	3	-1
	LU	3	-49	5	3	937	-622	0	5	16	39	-277	1	37	10	-144
	NL	7	-199	8	4	99	-1161	0	10	1	0	-875	0	18	21	-518
	PO	0	1	0	0	41	7	0	3	11	0	0	447	370	0	4
	ES	0	0	0	0	153	10	0	2	36	0	-1	220	1391	0	1
	SE	0	-3	29	20	0	24	0	1	0	0	-8	0	0	140	-4
	UK	0	-35	9	2	65	-137	0	26	1	0	-101	0	7	12	-1127
	EU*	10	-7	5	9	98	-17	14	2	57	0	-24	16	88	18	-41

Note: The table gives the mean reduction in AOT40 (ppb.h) in each receptor country (along left-hand side) caused by a 40% NOx emission reduction in each source country (along top) * denotes average over all land-areas of Europe.

Table 6.7. Country-to-country matrix for 40% VOC Reduction, AOT40-crops

		Source-country														
		AT	BE	DK	FI	FR	DE	GR	IE	IT	LU	NL	PO	ES	SE	UK
Receptor-country	AT	95	27	6	2	128	371	1	1	266	1	33	0	5	7	87
	BE	23	344	10	1	404	894	0	6	15	7	272	0	10	11	595
	DK	3	42	70	4	57	230	0	5	2	0	67	0	0	39	306
	FI	0	2	3	2	3	13	0	0	0	0	4	0	0	8	32
	FR	13	102	5	0	473	426	0	4	79	3	104	3	76	4	300
	DE	32	103	17	2	224	910	0	3	46	4	139	0	6	13	263
	GR	6	2	1	0	14	38	174	0	104	0	4	0	1	1	11
	IE	1	26	3	0	76	98	0	23	1	0	42	0	5	2	484
	IT	32	15	3	0	178	137	5	0	964	0	18	1	18	3	46
	LU	32	233	9	1	427	1143	0	4	29	19	208	0	14	8	325
	NL	19	235	17	1	256	908	0	7	5	3	354	0	6	14	653
	PO	1	6	0	0	54	24	0	1	7	0	6	361	331	0	41
	ES	2	15	0	0	98	55	0	1	19	0	15	96	389	0	40
	SE	0	6	12	2	9	48	0	1	0	0	13	0	0	21	64
	UK	3	56	6	0	133	219	0	11	3	1	80	0	3	6	919
	EU*	7	16	4	1	58	108	5	1	54	0	21	8	29	5	73

Note: The table gives the mean reduction in AOT40 (ppb.h) in each receptor country (along left-hand side) caused by a 40% VOC emission reduction in each source country (along top); * denotes average over all land-areas of Europe.

6.8 Summary and Conclusions

Table 6.8. Country-to-country matrix for 40% NO_x Reduction, AOT60

	\multicolumn{13}{c	}{Source-country}
		AT	BE	DK	FI	FR	DE	GR	IE	IT	LU	NL	PO	ES	SE	UK
Receptor-country	AT	92	0	0	0	23	82	0	0	80	0	-2	0	1	0	-3
	BE	3	31	1	0	414	-175	0	0	2	14	-127	0	20	4	-104
	DK	2	-7	41	0	18	85	0	0	0	0	-17	0	0	7	-32
	FI	0	0	1	1	0	1	0	0	0	0	0	0	0	2	0
	FR	4	5	1	0	446	60	0	0	30	4	-20	1	23	2	-38
	DE	25	-6	2	0	138	310	0	0	15	4	-39	0	4	2	-14
	GR	0	0	0	0	1	1	41	0	28	0	0	0	0	0	0
	IE	0	0	0	0	14	-6	0	19	0	0	-7	0	1	0	-6
	IT	11	0	0	0	59	12	1	0	378	0	0	0	7	0	0
	LU	1	62	2	0	583	290	0	0	12	58	-43	0	16	5	-3
	NL	5	-25	1	0	174	-327	0	0	-3	2	-157	0	12	3	-106
	PO	0	0	0	0	7	2	0	0	0	0	0	92	54	0	1
	ES	0	0	0	0	30	5	0	0	3	0	0	28	145	0	0
	SE	0	0	5	0	1	14	0	0	0	0	-2	0	0	9	-3
	UK	0	-13	1	0	54	-31	0	1	0	0	-31	0	1	1	-189
	EU*	4	0	0	0	36	18	0	0	19	0	-4	2	10	1	-7

Note: The table gives the mean reduction in AOT60 (ppb.h) in each receptor country (along left-hand side) caused by a 40% NOx emission reduction in each source country (along top); * denotes average over all land-areas of Europe.

Table 6.9. Country-to-country matrix for 40% VOC Reduction, AOT60

	\multicolumn{13}{c	}{Source-country}
		AT	BE	DK	FI	FR	DE	GR	IE	IT	LU	NL	PO	ES	SE	UK
Receptor-country	AT	4	1	0	0	4	17	0	0	9	0	1	0	0	0	2
	BE	8	84	2	0	73	349	0	0	4	1	72	0	1	3	127
	DK	2	13	1	0	17	60	0	0	2	0	17	0	0	0	35
	FI	0	0	0	0	0	1	0	0	0	0	0	0	0	0	1
	FR	2	15	0	0	25	82	0	0	5	0	14	0	0	0	43
	DE	6	23	0	0	40	155	0	0	10	1	23	0	1	1	23
	GR	0	0	0	0	0	1	24	0	4	0	0	0	0	0	0
	IE	0	5	0	0	6	28	0	0	0	0	5	0	0	0	57
	IT	1	0	0	0	6	8	0	0	104	0	0	0	0	0	2
	LU	9	33	1	0	66	304	0	0	12	2	37	0	2	2	42
	NL	5	77	2	0	84	338	0	0	6	1	87	0	2	2	133
	PO	0	0	0	0	3	2	0	0	0	0	0	64	28	0	1
	ES	0	0	0	0	1	2	0	0	0	0	0	6	9	0	1
	SE	0	1	0	0	1	8	0	0	1	0	1	0	0	0	2
	UK	1	20	0	0	30	87	0	0	1	0	25	0	0	0	177
	EU*	1	3	0	0	5	19	0	0	5	0	3	0	0	0	9

Note: The table gives the mean reduction in AOT60 (ppb.h) in each receptor country (along left-hand side) caused by a 40% VOC emission reduction in each source country (along top); * denotes average over all land-areas of Europe.

6.9 References

Amann M et al (1999), Final report on work done under EU acidification and ozone strategy. International Institute for Applied Systems Analysis (IIASA), Laxenburg, Austria

Andersson-Sköld Y and Simpson D (1999), Comparison of the chemical schemes of the EMEP MSC-W and the IVL photochemical trajectory models. In: *Atmospheric Environment*, 33, 1111-1129

DeMore, W.B. et al. (1997) Chemical kinetics and photochemical data for use in stratospheric modelling, JPL Publication 97-4.

Derwent, R.G., Jenkin, M.E., Saunders, S.M. (1996) Photochemical ozone creation potentialsfor a large number of reactive hydrocarbons under European conditions, *Atmos. Environ.*, 30, 189-200.

Eliassen, A., Hov, O., Isaksen, I.S.A., Saltbones, J. and Stordal, F. (1982) A Lagrangian long-range transport model with atmospheric boundary layer chemistry, *J.Appl. Met.*, 21, 1645-1661.

EMEP MSC-W (1998) Transboundary photooxidant air pollution in Europe. Calculations of tropospheric ozone and comparison with observations. Norwegian Meteorological Institute, Oslo, Norway

Fuhrer, J., Skärby, L., and Ashmore, M.A.(1997) Critical levels for ozone effects on vegetation in Europe, 1997, *Environ.Poll.*, pp 91-96.

Friedrich, R. (1993) Generation of time-dependent emission data, In P. Borrel et al., editor, Transport and Transformation of Pollutants in the Troposphere, Proceedings EUROTRAC symposium 1992, pages 255-268

Friedrich, R., Reis, S., Voehringer, F., Simpson, D., Moussiopoulos, N., Salmons, R., and Papameletiou, D. (1998) Efficient ozone abatement strategies in Europe, in Transport and Chemical Transformation of Pollutants in the Troposphere, Proceedings EUROTRAC symposium 1998.

Hass, H., Builtjes, P.J.H., Simpson, D., and Stern, R. (1997) Comparison of model results obtained with several European regional air quality models, *Atmos. Environ.*, 31, No. 19, 3259-3279

Jenkin, M.E., Saunders, S.M. and Pilling, M.J. (1997) The tropospheric degredation of volatile organic compounds: a protocol for mechanism development, *Atmos.Environ.*, 31, 81-104.

Jonson, J.E., Jakobsen, H.A., and Berge, E. (1997) Status of the development of the regional scale photo-chemical multi-layer eulerian model, Norwegian Meteorological Institute, EMEP MSC-W Note 2/97

Jonson, J.E., Tarrason, L., Sundet, J., Berntsen, T., and Unger, S. (1998) The Eulerian 3-D oxidant model: status and evaluation for summer 1996 results and case-studies, In EMEP MSC-W Report 2/98, Part II

Jonson, J.E., Tarrason, L., and Sundet, J. (1999) Calculation of ozone and other pollutants for the summer 1996, *Env. Manage. and Health, in press*

Kuhn, M., Builtjes, P.J.H., Poppe, D., Simpson, D., Stockwell, W.R., Andersson-Sköld, Y., Baart, A., Das, M., Fiedler, F., Hov, O., Kirchner, F., Makar, P.A., Milford, J.B., Roemer, M.G.M., Ruhnke, R., Strand, A., Vogel, B., and Vogel, H. (1998) Intercomparison of the gas-phase chemistry in several chemistry and transport models, *Atmos. Environ.*, 32, No. 4, 693-709

LAT (1997) Laboratory of Applied Thermodynamicy, University of Thessaloniki, Methodologies for Estimating Air Pollutant Emissions from Transport – Road Traffic Composition, Task 2.2 / Deliverable 16, LAT Report No. 9718, Thessaloniki, August 1997

LAT (1998) Laboratory of Applied Thermodynamicy, University of Thessaloniki, Methodologies for Estimating Air Pollutant Emissions from Transport – Emission Factors

6.9 References

and Traffic Characteristics Data Set, Deliverable 21, LAT Report No. 9802 Thessaloniki, January 1998

Middleton, P., Stockwell, W.R. and Carter, W.P.L. (1990) Aggregation and analysis of volatile organic compound emissions for regional modeling, *Atmos. Environ.*, 24A, No. 5, 1107-1133.

Moussiopoulos, N., Sahm P., Tourlou, M., Friedrich, R., Simpson, D., Lutz, M. (1999) Development and evaluation of ozone abatement strategies on urban scale air quality, 8th International Symposium Transport and Air Pollution, Graz, Austria.

Pleim, J. E. and Ching, J.K.S. (1993) Interpretative analysis of observed and modeled mesoscale ozone photochemistry in area with numerous point sources, Atmos.Environ., 27A, 999-1017.

Reis, S., Simpson, D., Friedrich, R., Jonson, J.E., Unger, S., and Obermeier, A. (1999) Road traffic emissions - predictions of future contributions to regional ozone levels in Europe, 8th International Symposium Transport and Air Pollution, Graz, Austria.

Sillman, S., Logan, A., and Wofsy, S.C. (1990) The sensitivity of ozone to nitrogen oxides and hydrocarbons in regional ozone episodes, *J. Geophys. Res.*, 20, 1837-1851.

Simpson, D. (1992a) Long period modelling of photochemical oxidants in Europe. Calculations for July 1985, *Atmos. Environ.*, 26A, No. 9, 1609-1634.Simpson, D.(1992b) Long period modelling of photochemical oxidants in Europe. A) Hydrocarbon reactivity and ozone formation in Europe. B) On the linearity of country-to-country ozone calculations in Europe, EMEP MSC-W Note 1/92, Norwegian Meteorological Institute, Oslo.

Simpson, D. (1993) Photochemical model calculations over Europe for two extended summer periods: 1985 and 1989. Model results and comparisons with observations, *Atmos. Environ.*, 27A, No. 6, 921-943

Simpson, D. (1995) Biogenic emissions in Europe 2: Implications for ozone control strategies, *J. Geophys. Res.*, 100, No. D11, 22891-22906.Simpson, D. and Andersson-Sköld, Y. (1997) Regional and local scale modelling of ozone in Europe:calculations for the EU Auto/Oil Programme. Norwegian Meteorological Institute, Research Report No. 47, Oslo, Norway.

Simpson, D. and Jonson, J.E. (1998) Comparison of Lagrangian and Eulerian models for the summer of 1996, In EMEP MSC-W Report 2/98, Part III, Transboundary photooxidant air pollution in Europe. Calculations of tropospheric ozone and comparison with observations. Norwegian Meteorological Institute, Oslo, Norway

Simpson, D. and Malik, S. (1996) Photochemical oxidant modelling and source-receptor relationships for ozone, In EMEP MSC-W Report 1/96, Norwgian Meteorological Institute, Oslo, Norway.

Simpson, D., Guenther, A., Hewitt, C.N., and Steinbrecher, R. (1995) Biogenic emissions in Europe 1. Estimates and uncertainties, *J. Geophys. Res.*, 100, No. D11, 22875-22890

Simpson, D., Olendrzynski, K., Semb, A., Storen, E., and Unger, S. (1997) Photochemical oxidant modelling in Europe: multi-annual modelling and source-receptor relationships, Norwegian Meteorological Institute, EMEP MSC-W Report 3/97.

Simpson, D., Altenstedt, J., and Hjellbrekke, A.G. (1998) The Lagrangian oxidant model: status and multi-annual evaluation, Part I in MSC-W (1998).

Solberg, S. , Dye, C. , Schmidbauer, N. , and Simpson, D. (1995) Evaluation of the VOC measurement programme within EMEP, Kjeller, Norway, Norwegian Institute for Air Research (NILU) Report 5/95.

UN-ECE (1997) Health-effects of Ozone and Nitrogen Oxides in an Integrated Assessment of Air Pollution, Institute for Environment and Health, University of Leicester, Leicester, UK.

7 Optimising Regional Ozone Strategies

Stefan Reis, David Simpson and Rainer Fr

7.1
Introduction

Chap. 6 has shown some effects of EU-wide emission controls on regional ozone concentrations in Europe. However, as discussed in detail within the following sections, an iterative approach is needed to optimise abatement strategies to reach a specific ozone threshold throughout the EU. And since the full EMEP photochemical oxidant model would need excessive amounts of processor time to be applied for e.g. several hundred optimisation steps, a simpler model (see *Sect. 7.1.1*) has to be used at this stage, allowing the investigation of a considerable number of scenarios.

In this chapter the possibility of finding cost-effective methods of distributing emission controls for NO_x and NMVOC across different countries within the EU is being explored. Such country-specific emission control options have been used for some years now. Indeed, recent UNECE and EU emission strategies have specified sets of country-specific emission reductions which are calculated to meet agreed environmental targets at least cost. Further, these environmental targets are spatially variable, especially for acidification problems, being based upon so-called critical thresholds, which represent levels of deposition or concentration above which environmental damage is believed to occur. The UN ECE 2nd sulphur Protocol represented the first move towards this more complex but cost-effective approach. Optimisation techniques (ASAM - ApSimon et al.1994; CASM - SEI, 1991; RAINS - Alcamo et al., 1990) were applied together with information on the cost of emission reductions in each country (abatement cost curves), EMEP calculations of country-to-grid depositions (Barrett and Sandnes, 1996), and maps of critical loads for sulphur (Posch et al., 1995). This so-called integrated assessment modelling resulted in a set of target emissions which varied from country-to-country. Similar techniques have recently been developed for ozone (Schöpp et al., 1999), using a parameterisation of EMEP MSC-W modelling results as the basis, and these were used as part of the EU *Acidification and Ozone Strategy* to set EU national Emissions Ceilings (Amann et al., 1999).

The main problem in optimising strategies for ozone lies in the treatment of NO_x emissions and their role in both acidification and ground-level ozone, both of great concern in Europe. Unfortunately, measures to reduce NO_x in different parts have quite different effects on acidification than on ozone al., 1994). In addition, emissions control of volatile organic (VOC) may in some regions be a more cost-effective approach ozone control than NO_x reduction. The EU has currently 15 countries, and 15 sets each of NO_x emissions, NMVOC emissions, NO_x-cost curves and NMVOC-cost curves (assuming independence of these), which should be used as a basis for the optimisation. Thus it is clear that even for a limited number of precursors and environmental problems, finding the most cost-effective solution among so many variables is a formidable task.

Within this study, an iterative methodology has been developed for addressing an optimisation problem for a non-linear relationship of two precursors of tropospheric ozone, NO_x and NMVOC. In principle, this methodology could be applied with presently available EMEP data to handle simultaneously the effects of SO_2, NO_x, VOC, and NH_3 on acidification, eutrophication, ozone-crop damage, ozone-forest damage, and ozone-health effects. In this case the methodology has been applied mainly to ozone, although with some attention payed to the linkage to acid deposition. This methodology along with some illustrative case-studies, has been presented in detail in Simpson and Eliassen (1997, 1999), but the main principles will be described briefly here.

The main improvements in the methodology since Simpson and Eliassen (1999) has been in the implementation of 'real' cost curves rather than the generic cost curves originally used. This improved model, which is now termed *OMEGA (Optimisation Model for Environmental Integrated Assessment)* will be discussed in *Sect. 7.2*. In *Sect. 7.2.2* the concept of 'optimum' solutions and differences between the iteration methodology and other optimisers is discussed briefly. In *Sect. 7.3* the methodology is applied to derive cost-effective strategies aimed at different targets. Having identified some promising strategies and illustrated the environmental improvements arising from these, *Sect. 7.4* shows the evaluation of optimisation results versus full EMEP model results, which is followed by an examination of the modelled exceedance statistics with different scenarios (*Sect. 7.5*).

7.2
The OMEGA Model

Outline

The optimisation procedure consists of typically hundreds of iteration steps, following the abatement cost curve (for each pollutant) of each country to find the least-cost option to achieve a preset target. At each step the emissions of all active countries are reduced by a small amount (a country is active as long as further emission reductions are possible for that country). At each step of the iteration, each country has reached a specific location on its abatement cost curve for NO_x and as well for NMVOC. This location gives a level of abatement costs and the related percentage of emissions abated. As the optimisation proceeds, every next incremental step of reduction becomes more costly for this country, the model always searching for the least-cost option to increase the *benefit* (i.e. reduced overall concentration of ozone). The *benefit* associated with a given environmental change may be defined in many ways of course. This can be illustrated with *Fig. 7.1*, using the example of AOT40. At the beginning of the iterations there is the initial field of AOT40 (AOT40(n=0)). After *n* iterations this field is updated to AOT40(*n*), which, however, still lies above a desired critical level CL_{AOT40}. A NO_x emission change from country *i* produces a change $\Delta AOT40 \equiv MM'$ in the calculated field, from AOT40(n) to AOT40(n+1).

Thus, the *benefit* could be any desired function of MM'. Most simply, the *benefit* assigned to emitter *i* could be defined as proportional to MM' ($\equiv \Delta AOT40$). An alternative approach used here is to define the benefit in terms of fractional gap closure, where 'gap' refers to the difference between current pollutant levels and the desired critical threshold (i.e. PN). Thus, here the benefit is set proportional to MM'/PN. The first of these options focuses on areas with the largest exceedances of the critical level, whereas the second aims at a more geographically even distribution of AOT40 reduction, thus favouring a balanced reduction of ambient concentrations in many grid cells against a comparatively large reduction in only one grid cell. In order to investigate the influence of the approach selected in OMEGA model calculations were conducted using either approach – closing relative gaps instead of reducing absolute concentrations – on a 50 % reduction of the initial total ozone concentration. While reaching the same target, the reductions of precursor emissions neccessary to achieve this differ considerably for NO_x. The gap closure approach leads to a more balanced distribution of reductions among the countries, reflecting the model attempting to reduce ozone concentrations for all grid cells above the threshold, instead of focussing on some grid cells with high ozone levels only. For NMVOC, the results for the gap closure approach does not show significant differences.

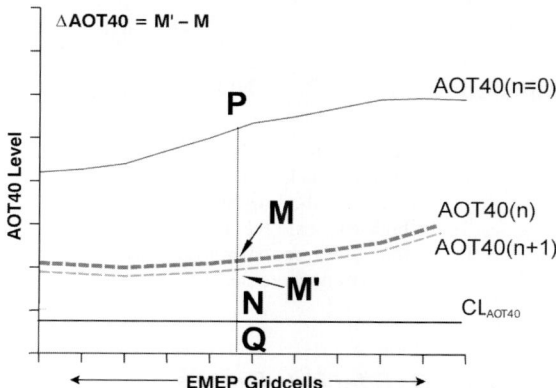

Fig. 7.1. Sketch of iteration procedure. After n iteration steps the field of AOT40 across some EMEP grid squares is reduced from its initial value AOT40(0) to AOT40(n). The further reduction from AOT40(n) to AOT40(n') is used as a measure of the "benefit".

The choice of defining the benefit depends on the targets set, e.g. the reduction of peak concentrations, or the reduction of overall exposure. Basically, any benefit definition can be implemented in OMEGA.

In a multi-effect scenario benefits for each effect must be weighed up against all other effects. This is accomplished by assigning weighting factors. With our example two-effect-problem (acidification, ozone) interest in acid deposition is given by a parameter w_{acid}, and interest in ozone by w_{ozone}, with $w_{acid} + w_{ozone} = 1$. One of the advantages of this approach is that it can easily be extended to cover the same treatment to any number of pollutants and environmental problems. For example, an interest in tropospheric ozone could be specified with W_{ozone} and an interest in eutrophication with $W_{eutroph}$, with the simple requirement that the sum of all weighting factors is one. Since the main objective of this work was to reduce tropospheric ozone, calculations have mainly been done with $w_{ozone} = 1.0$ (i.e. $w_{acid} = 0$). To investigate the influence of acidification on the requirements for NO_x emission abatement, some additional model calculations have been conducted with $w_{ozone} = w_{acid} = 0.5$. Thus, the formulation of the iteration methodology is straightforward. Indeed most of the complications presented in Simpson and Eliassen (1997, 1999) are associated with the need to sum over all emitters and receivers, and the normalisation of benefits.

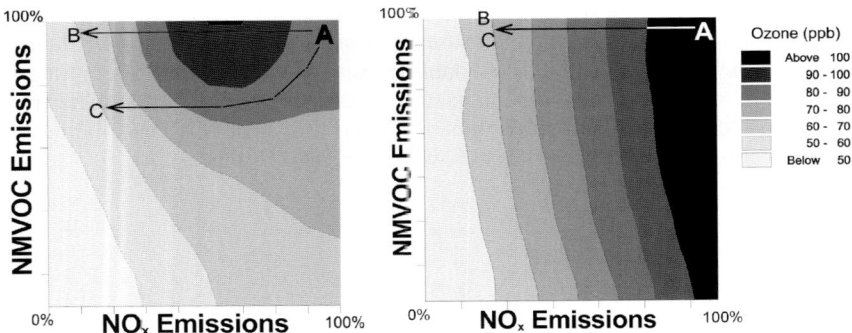

Fig. 7.2. Illustrative ozone isopleth diagrams for low-NO_x regions (right) and high-NO_x regions (left). The arrows illustrate the possible behaviours of non-linear optimisers (A→B) as compared to the iteration model (A→C). In the low-NO_x case both optimisers will choose the same route to reduced ozone. In the high-NO_x case the non-linear optimiser may see through the 'hill' and suggest point B as the optimum solution, whereas the iteration model will skirt around the hill to get to point C, preventing any ozone increase.

Treatment of Negative NO_x Effects (γ-Factors)

An important problem for optimisation is that NO_x emission reduction sometimes leads to increased ozone in some areas (Sillman et al., 1990; Simpson, 1995). Such behaviour may prevent an optimisation programme from recommending NO_x emission reductions from affected countries, and in extremis might lead to a mathematical request for higher emissions! Such behaviour is clearly not politically acceptable, and is certainly environmentally unacceptable given the range of problems which are associated with NO_x emissions.

To illustrate, *Fig. 7.2* shows possible isopleths of ozone concentrations for both a low-NOx situation and a high-NO_x situation. In the low-NO_x case, ozone is steadily reduced with reductions in either NO_x or VOC emissions. In the high-NO_x case ozone is also reduced with VOC reductions, but increases at first with reduced NOx emissions. The low-NO_x situation is typical over most of Europe, but the high-NO_x case occurs for Belgium, Germany, Luxembourg, Netherlands and the United Kingdom in the simulations carried out here.

Non-linear methods exist (e.g. Schöpp et al., 1999, Heyes et al., 1996) of seeing beyond the 'hill' when searching for the optimum ozone strategy. However, a problem with this approach is that it allows ozone to increase in the first years of the control policy (see *Sect. 7.2.2*). Our approach is to design a method which prevents ozone increasing at any stage of the emission control implementation. *Fig. 7.2* indicates that this can be accomplished by simply demanding that a reduction in NO_x emissions is accompanied by a simultaneous reduction in VOC emissions. The amount of VOC reduction is given by a factor γ, where

$$\Delta E_{voc} = \gamma \times \Delta E_{nox} \qquad (E\ 7.1)$$

There are obviously many ways in which γ can be determined. The strictest would be to demand that ozone (or AOT40) could not increase in any grid square, but the amount of VOC reduction required to achieve this everywhere would be considerable: γ values of more than 10 would be required in some cases. An alternative approach used here is to ask that the average AOT40 over any country does not increase. The γ values thus obtained are: 1.4 -Belgium; 0.96 - Germany; 0.45 - Italy; 0.34 - Luxembourg; 1.3 - Netherlands; 1.4 - United Kingdom.

7.2.1
Inputs to the Optimisation

The basic input data for the OMEGA model consists of national emissions data of NO_x and VOC, cost curves for NOx and VOC per country, critical or target levels data which specifies the pollution level of acid deposition or ozone deemed acceptable to prevent significant environmental effects in each grid square, initial calculations of base-case pollutant loads, and finally source-receptor matrices relating changes in pollution to emissions from each country.

Emissions

Emissions are taken from data generated for the 2010 trend scenario (*Chaps. 3, 5*) or for non-EU countries from the officially submitted (to EMEP) *Current Reduction Plans* (CRPs) or EMEP estimates for 2010 in case no officially submitted data is available (Olendrzynski 1997).

Costs & Abatement Cost Curves

It is clear that the results of the optimisation model are determined to a large extent by the costs associated with different emission measures. Cost estimates should be entered into the calculations on a country-by-country basis for the effects of each emission measure. Further, a measure or technology which reduces both NO_x and VOC should have one cost associated with the reduction of two pollutants. However, this latter consideration will become less important in coming years. Passenger cars are the only major source category with both significant NO_x and VOC emissions. As most car fleets will have turned over to catalyst cars by the year 2010, this source category should no longer be major. Other sources categories tend to be dominated by one pollutant or the other; e.g. combustion sources and heavy-duty traffic by NO_x and solvents and petroleum refining by VOC.

The important feature is that costs increase only gradually up to a certain point, often referred to as the 'knuckle' point, after which emission reduction becomes dramatically more expensive, showing the behaviour similar to an exponentially sloping curve. These abatement cost curves are of major importance, as they drive the selection of reductions for each country and thus are one of the main parameters determining how an optimised solution will look like.

For this study, special care was taken to accurately assess the detailed costs for each available abatement measure, split into the main cost features *investment* and *labour costs* (for the first installation of a measure), *operating and maintenance* costs (depending on utilisation and operation of a plant or vehicle), *energy costs* (e.g. for increased energy demand due to additional equipment) and finally *savings*, i.e. negative costs (e.g. from recovered solvents in closed systems). It is vital to assess these costs as exactly as possible, because they are the main drivers of the optimisation procedure later on. The detailed description of measures and related costs can be found in *Chap. 4*. In order to determine, which would be the costs of the abatement measures implemented and to distinguish them from costs occurring because of other developments, the approach of additional costs was taken, i.e. the difference of e.g. the costs of a combustion plant with a low-NO_x-burner and the costs of a plant with a standard burner. Investment and labour costs for the first installation were annualised over the average lifetime of the equipment, variable costs (O&M etc.) were collected on an annual basis, giving a total annual additional costs per measure as input to the cost curve calculation.

By using activity data for each source group and assessing the implementation degree and total abatement potential, the total costs of abatement could be transferred into specific abatement costs. as unit costs in ECU tonne of pollutant abated, which is the only way to efficiently compare measures applying to different sources.

While this is comparably easy for measures only reducing one pollutant, e.g. energy sector measures reducing NO_x, it is necessary to distribute the costs of measures that are responsible for the reduction of two or more relevant pollutants. This is especially the case for road transport measures, which consist of technology bundles to achieve compliance with EURO emission standards. These measures usually reduce NO_x and NMVOC, sometimes even CO at the same time.

For this study, costs were attributed to NO_x and NMVOC proportional to the marginal abatement costs at the optimum solution (this requires an iterative procedure).

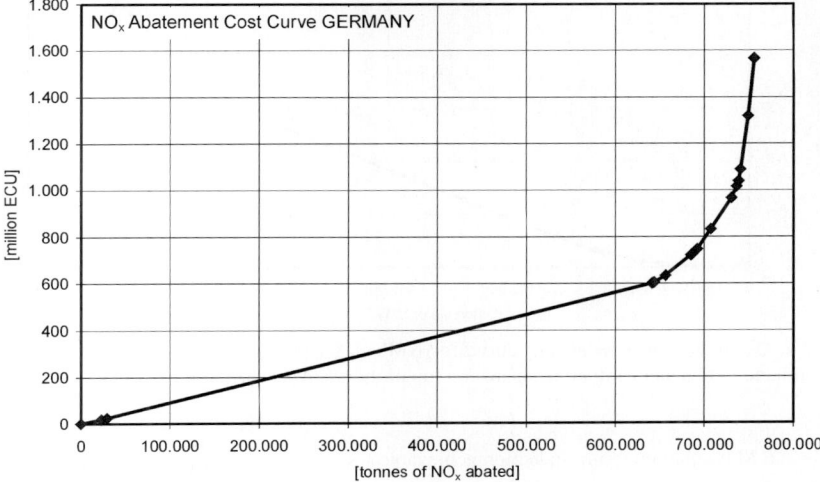

Fig. 7.3. Exemplary abatement cost curve for NO_x, Germany (accumulated total costs vs. accumulated total abatement, all measures on top of the Trend Scenario for 2010)

Table 7.1. Maximum emission reduction achievable for NO_x with measures included in the cost curves

NO_x	Emissions remaining (ktonnes)	Reduction from Base Case 1990	Reduction from Trend 2010	Total costs at maximum reduction (MECU)	Share of total costs	Mean unit costs (kECU/t)
Austria	101	-56%	-40%	127	1.3%	1.8
Belgium	119	-65%	-40%	304	3.0%	3.8
Denmark	107	-61%	-32%	256	2.6%	5.0
Finland	133	-51%	-50%	474	4.7%	3.5
France	615	-61%	-42%	960	9.6%	2.2
Germany	1 185	-60%	-39%	1566	15.7%	2.1
Greece	375	-31%	-28%	358	3.6%	2.4
Ireland	49	-57%	-64%	286	2.9%	3.3
Italy	855	-58%	-37%	1 409	14.1%	2.8
Luxembourg	14	-41%	-12%	9	0.1%	4.3
Netherlands	296	-47%	-28%	184	1.8%	1.6
Portugal	102	-53%	-53%	354	3.5%	3.0
Spain	657	-47%	-39%	1 223	12.2%	2.9
Sweden	218	-37%	-32%	182	1.8%	1.8
United Kingdom	830	-70%	-51%	2 308	23.1%	2.7
EU15	**5 655**	**-58%**	**-41%**	**10 001**		**2.6**

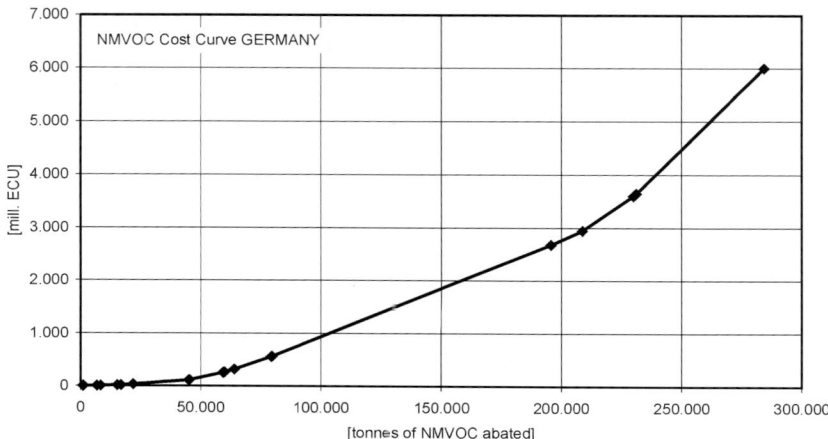

Fig. 7.4. Exemplary abatement cost curve for NMVOC, Germany (accumulated total costs vs. accumulated total abatement, all measures on top of the trend scenario for 2010)

Table 7.2. Maximum emission reduction achievable for NMVOC with measures included in the cost curves

NMVOC	Emissions remaining (ktonnes)	Reduction from Base Case 1990	Reduction from Trend 2010	Total costs at maximum reduction (MECU)	Share of total costs	Mean unit costs (kECU/t)
Austria	288	-31%	-4%	523	1.6%	18.9
Belgium	167	-54%	-14%	650	1.9%	31.9
Denmark	61	-64%	-30%	501	1.5%	19.2
Finland	97	-41%	-18%	486	1.5%	23.3
France	1 108	-54%	-25%	7 288	21.8%	20.2
Germany	1 270	-57%	-20%	6 053	18.1%	21.2
Greece	198	-39%	-30%	789	2.4%	9.3
Ireland	112	-38%	-16%	347	1.0%	17.2
Italy	1 394	-42%	-17%	6 046	18.0%	21.8
Luxembourg	8	-55%	-26%	53	0.2%	21.1
Netherlands	300	-34%	-9%	707	2.1%	23.9
Portugal	138	-33%	-12%	665	2.0%	35.6
Spain	717	-36%	-19%	5 328	15.9%	32.3
Sweden	288	-36%	-11%	1 025	3.1%	28.2
United Kingdom	1 408	-46%	-13%	3 040	9.1%	17.6
EU15	**7 556**	**-47%**	**-17%**	**33 502**		**21.6**

Taking into account interdependency and exclusiveness of some measures, they were ranked according to their unit abatement costs and transformed into piecewise linear abatement cost curves as shown in *Figs. 7.3* and *7.4*. These curves haven then been implemented into OMEGA, giving a detailed abatement cost curve for NO_x and NMVOC for each EU15 Member State. *Tables 7.1* and *7.2* show the maximum reductions that could be achieved by implementing all measures included in the abatement cost curves, along with their related total and unit costs.

Critical Thresholds

The critical levels for acid deposition represent the amount of sulphur and/or nitrogen-deposition which each grid square can tolerate if the aim is to protect 95 % of the ecosystems, both with respect to acidity and eutrophication. The critical level maps used here were generated by the *Co-ordinating Centre for Effects* (Posch et al. 1995), update of February 1997, and assume a fixed sulphur deposition equivalent to implementation of the 2nd Sulphur Protocol (year 2010).

For ozone, critical thresholds are defined in terms of accumulated exceedance statistics. For vegetation AOT40 is used (Fuhrer et al., 1997, see *Chap. 2*). For crops, the growing season for modelling purposes is defined as the 3 months May-July, and this statistic is referred to as AOT40c. The critical threshold for AOT40c is currently set to 3000 ppb hours.

For health, the UN-ECE workshop on 'health effects of ozone and nitrogen oxides in an integrated assessment of air pollution' (UN-ECE, 1997) agreed that a simple statistic such as AOT60 could be used as a preliminary indication of ozone levels above the recommended WHO guideline for integrated assessment modelling purposes. AOT60 is defined in an analogous manner to AOT40 above, but no critical threshold is defined for AOT60 as all ozone over 60 ppb is thought to be damaging for human health. (cf. *Chap. 2*)

Base-Case Exceedances

Base-case (year 2010) fields of ozone and N-deposition for the year 2010 are taken directly from EMEP acid deposition and ozone model calculations, as described in *Chap. 6*.

Source-Receptor Relationships

Source-receptor matrices relate changes in pollution at a receptor (country or grid square) j to the emissions from an emitter country i. The source-receptor (S-R) relationships used here have been presented in EMEP Report 1/96 for acid deposition (Barrett and Sandnes 1996), and in *Chap. 6* for ozone (AOT40c and AOT60).

In order to deal with the variable nature of ozone S-R relationships, two sets of matrices have been used, one derived from 2010 CRP emission levels and the second from a hypothetical low-emission situation (40 % of 2010 emissions). For any given country the S-R relationship used in the iteration model is calculated by interpolation between these two base S-R matrices, depending on the level of NO_x emissions at each iteration step (see Simpson and Eliassen 1997 for more details).

7.2.2
An Optimised Solution?

With all optimisation models there is a possibility that the methodology might not find the absolute minimum cost solution, but rather might find a local-minimum in cost-space. Sophisticated non-linear optimisation techniques (e.g. RAINS, Alcamo et al. 1990, Heyes et al. 1996) are designed to avoid such local minima. The iteration method presented here is designed to be transparent and flexible, rather than mathematically rigorous. There is thus no guarantee that in the end, the solution of the OMEGA model is *the* global least-cost solution. However, this apparent difficulty is probably not important, for the following reasons:

Global Optimisation Might Lead to Adverse Short-Term Effects.

A relevant example is the situation where ozone first increases with reduced NO_x emissions, but then decreases after NO_x emissions have been substantially reduced (the ozone *hill* problem, see *Fig. 7.2*). A global optimisation may see the other side of this ozone *hill* and recommend very drastic NO_x emissions reductions, possibly with little consideration of VOC reductions. In other words global optimisation may choose to ignore the hill and settle in a least-cost valley on the other side. The iteration method would not allow any NO_x reductions in this area without sufficient VOC reductions to prevent ozone increases. In other words the iteration method would skirt around the hill.

Although the non-linear model may have found the global least-cost solution correctly, following such a procedure might entail increased ozone in the years before full implementation of control measures took effect. The iteration model may not achieve the global-minima of costs by the given target year, but would ensure that ozone levels did not increase before the target year, and that measures were taken in a cost-effective and environmentally beneficial way at all points of the iteration.

Global Optimisation Might Lead to Adverse Long-Term Effects

Even the most ardent model enthusiasts must admit that it is dangerous to rely on the results of any model to be correct for situations 20 – 40 years ahead, when emission reductions of order 80 % are being considered. Thus, the possibility exists that features such as the ozone hill discussed above may be artefacts, due to unavoidable uncertainties connected with, for example, biogenic or man-made emissions, model formulation, or even basic scientific understanding. Thus a global optimisation which relies upon the model results to be accurate at all extremes of its prediction may give a worse answer than a simpler iterative approach which proceeds in small steps with an emphasis on safe strategies at all emission levels.

7.3 OMEGA Model Results

As was described in *Sect. 7.2*, the output of OMEGA can be described by the following equation (E 7.2):

$$Output_\Omega[\Delta E_{NOx}, \Delta E_{VOC}, C_{NOx}, C_{VOC}, Mc_{O_3}] = f(GC, wt, E_{0_{NOx}}, E_{0_{VOC}}, CC_{NOx}, CC_{VOC})$$

with

ΔE_{NOx}, ΔE_{NMVOC}	=	necessary emission reductions for NO_x, NMVOC
C_{NOx}, C_{NMVOC}	=	abatement costs for NO_x, NMVOC at the reduction level
Mc_{O3}	=	map of resulting ozone concentrations at 150×150 km
GC	=	Gap Closure set for each specific threshold $AOT40_{crops/forests}$ or AOT60
wt	=	weight attributed to acidification *(zero, if only ozone is modelled)*
E_{0NOx}, E_{0NMVOC}	=	emission basis for NO_x and NMVOC
CC_{NOx}, CC_{NMVOC}	=	abatement cost curves for NO_x and NMVOC

OMEGA has been used to evaluate an extensive set of emissions scenarios, ranging from attempts at modest gap-closure to costly attempts at 90 % gap-closure. *Plates 10-12* show the results of some of these gap-closure attempts forAOT60, $AOT40_{crops}$, and $AOT40_{forests}$.

After intensive testing and evaluation work, a set of targets for optimisation was defined. The cost curves showed varying reduction potentials for the different EU15 countries, reaching a maximum reduction for NO_x at between 13 % and 55 % and for NMVOC at between 4 % and 22 % on top of the 2010 trend scenario, these parameters limiting the achievable reduction of ozone concentrations. Since the feasible number of scenarios for macroeconomic modelling and equity assessment was limited, a set of realistically achievable targets was chosen *(Table 7.3)* and the full set of costs per country as well as emission reductions was calculated by OMEGA.

Table 7.3. Scenarios selected for assessment

	Gap Closure		
W_{ozone} : W_{acid}	AOT 40c	AOT 40f	AOT 60
1.0 : 0	15 %	33 %	33 %
0.5 : 0.5	–	–	33 %

The results for these scenarios (see *Tabs 7.1* and *7.2*) indicated that even by implementing all abatement measures included in the cost curves, gap closures achievable (relative to the situation in the Trend Scenario) ranged between 15 % and 35 % for different AOT. Thus, in a second step, OMEGA was applied to determine how far emissions would have to be reduced to achieve a significant Gap Closure for AOT 40 and AOT60. For this second step, generic abatement cost curves had to be used, reflecting increasing unit costs of abatement at each step, but allowing for a total abatement of emissions. Even though this is only possible in theory, it provides valuable information about which target levels for ground-level ozone are feasible to be achieved at all.

For AOT60, *Plate 11* shows that very large reductions in AOT60 are indeed possible (up to about 89 % GC), resulting in zero AOT60 over nearly all of the EU, along with significant benefits to non-EU countries.

Table 7.4. Resulting NO_x emissions from EU15 countries for each optimisation scenario

NO_x (in ktonnes)	Trend 2010	Optimisation scenarios			
		w_{ozone}=1.0, w_{acid}=0.0			$w_{ozone} = w_{acid} = 0.5$
		AOT60 33%	AOT40c 15%	AOT40f 33%	AOT60 33%
Austria	169	98	100	101	98
Belgium	199	185	193	195	172
Denmark	158	106	106	138	106
Finland	268	132	176	227	132
France	1 055	604	617	631	605
Germany	1 941	1 732	1 836	1 881	1 687
Greece	523	373	374	387	373
Ireland	137	48	51	80	48
Italy	1 356	847	862	906	847
Luxembourg	16	14	14	15	14
Netherlands	410	396	402	406	392
Portugal	219	99	117	153	99
Spain	1 077	652	759	840	653
Sweden	320	216	216	251	216
United Kingdom	1 689	1 630	1 640	1 664	1 579
Total EU15	**9 538**	**7 132**	**7 464**	**7 874**	**7 023**
change rel. to 2010		*- 25 %*	*- 22 %*	*- 17 %*	*- 26 %*
change rel. to 1990	*- 30 %*	*- 47 %*	*- 45 %*	*- 42 %*	*- 48 %*

For $AOT40_{crops}$, significant reductions are again possible (*Plate 12*), but at about 68 % GC a limit is reached with some areas of the EU are still exceeding the 3 ppm.h critical threshold. The location of the exceedances indicates, however, that emissions from non-EU countries, which have not been included in the optimisation, might be the cause. Finally, for $AOT40_{forests}$ (*Plate 13*) the critical threshold of 10 ppm.h is achieved in most areas with 70 % GC, and again levels of AOT40 are reduced significantly from the Trend Scenario 2010 with even 33 % or 50 % GC scenarios. Here, a maximum GC of even 97 % could theoretically be achieved. Apart from that, these runs show quite well, which order of a magnitude of emission reductions would be necessary to achieve close to zero exceedances of ozone thresholds (see *Chap. 11*). Though, for gap closures of 50% and more, it would be necessary to reduce emissions of individual countries to almost zero, which is not likely to be feasible.

7.3 OMEGA Model Results

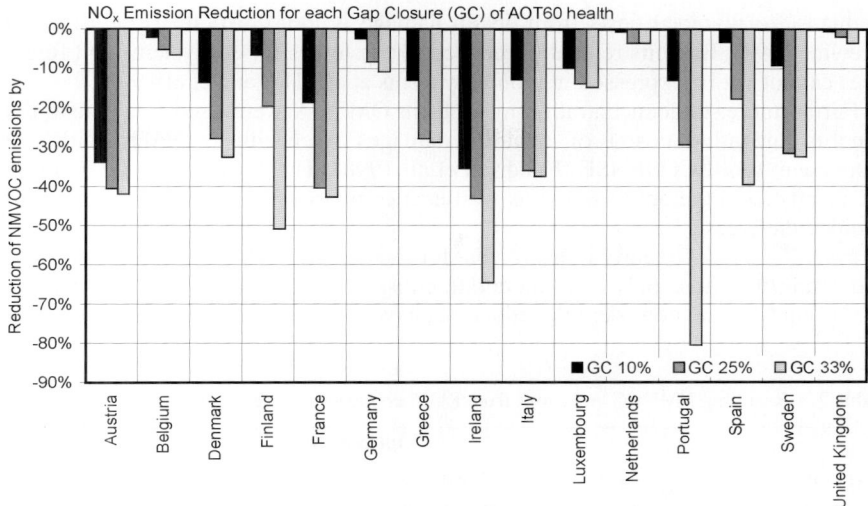

Fig. 7.5. Reduction of NO$_x$ emissions necessary for each country to achieve given GC of AOT60 health (with w$_{ozone}$=1.0, emission reductions relative to 2010 trend emissions)

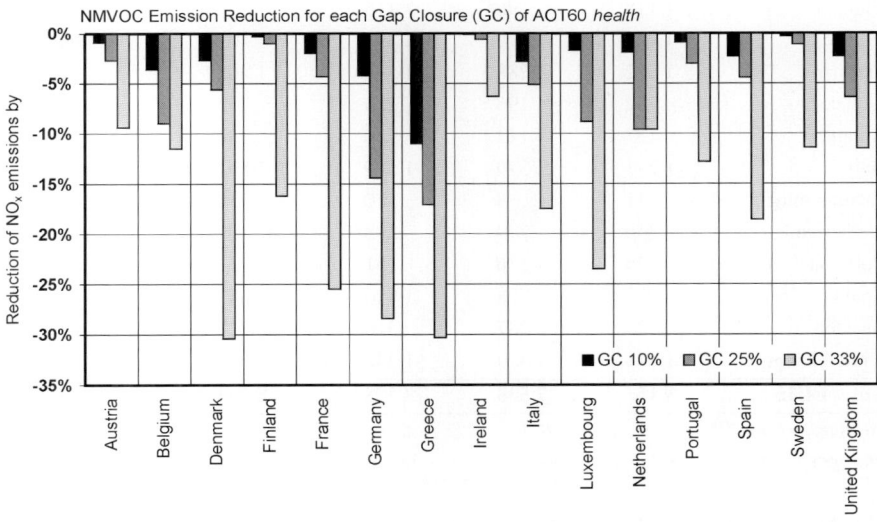

Fig. 7.6. Reduction of NMVOC emissions ecessary for each country to achieve given GC of AOT60 health (With w$_{ozone}$=1.0, emission reductions relative to 2010 trend emissions)

Along with the scenarios listed in *Table 7.3*, a 10 % and 25 % Gap Closure was calculated for AOT 60 to assess the relative changes of abatement costs and benefits. In the case of tropospheric ozone, the *benefits* are threefold: while increases in crop-yield can be expressed in monetary terms rather easily, the evaluation of avoided costs from health damage is more difficult (Friedrich and Krewitt, 1999). Benefits related to reduced ozone levels for ecosystems and forest trees cannot yet be expressed in monetary terms at all (Holland et al 1998).

Furthermore, the concentration maps from OMEGA were used to evaluate the benefits (negative costs) of avoided damages to health or crops with the assessment tool ECOSENSE (Friedrich et al. 1998). The *Figs. 7.5 to 7.9* describe the results from the optimisation as well as a comparison of abatement costs with avoided damages.

Figs. 7.5 and *7.6* clearly indicate, that for most countries a Gap Closure of 10 % and even 25 % does only need moderate emission reductions, while to achieve a 33 % Gap Closure, considerable reductions have to be undertaken.

Table 7.5. Resulting NMVOC emissions from EU15 countries for each optimisation scenario

NMVOC (in ktonnes)	Trend 2010	Optimisation scenarios			
		$w_{ozone}=1.0, w_{acid}=0.0$			$w_{ozone} = w_{acid} = 0.5$
		AOT60 33%	AOT40c 15%	AOT40f 33%	AOT60 33%
Austria	299	271	290	291	271
Belgium	194	172	175	177	172
Denmark	88	61	78	83	61
Finland	118	96	115	117	96
France	1 469	1 096	1 393	1 403	1 096
Germany	1 582	1 288	1 338	1 408	1 291
Greece	282	196	233	242	196
Ireland	133	112	130	132	112
Italy	1 671	1 379	1 589	1 591	1 379
Luxembourg	11	8	10	10	8
Netherlands	330	299	298	306	298
Portugal	156	136	151	152	136
Spain	882	708	840	843	708
Sweden	324	287	316	320	287
United Kingdom	1 615	1 429	1 440	1 519	1 431
Total EU15	**9 155**	**7 538**	**8 398**	**8 594**	**7 542**
changes rel. to 2010		*- 18 %*	*- 8 %*	*- 6 %*	*- 18 %*
changes rel. to 1990	*- 36 %*	*- 44 %*	*- 38 %*	*- 37 %*	*- 44 %*

The Netherlands and the United Kingdom are special cases, which – as long as acidification and thus NO_x-induced damages alone are not taken into account – do not have to reduce their NO_x emissions to a significant extent. With acidification included (i.e. $w_{acid} = 0.5, w_{ozone} = 0.5$) the concentration maps (*Plates 14 – 16 bottom*) show only slight differences (see *Sect.. 11.3* as well) due to forcing countries like the United Kingdom or the Netherlands to reduce their NO_x emissions to a somewhat greater extent. For NMVOC, the situation is more balanced, though it can be noted that the 33 % Gap Closure requires the full reduction potential in almost all countries, being close to the maximum reduction achievable with the set of abatement measures taken into account.

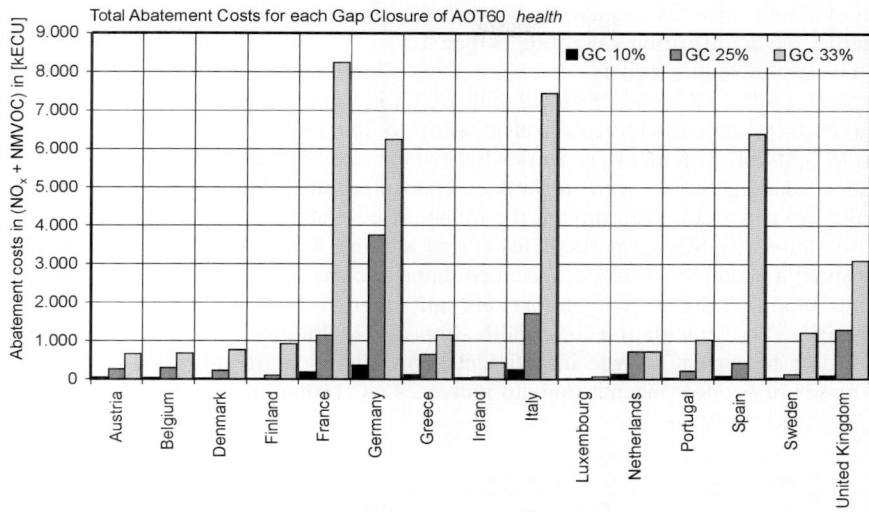

Fig. 7.7. Total abatement costs (NO$_x$ + NMVOC) to achieve the given Gap Closures of AOT 60 health (costs on top of the 2010 trend scenario)

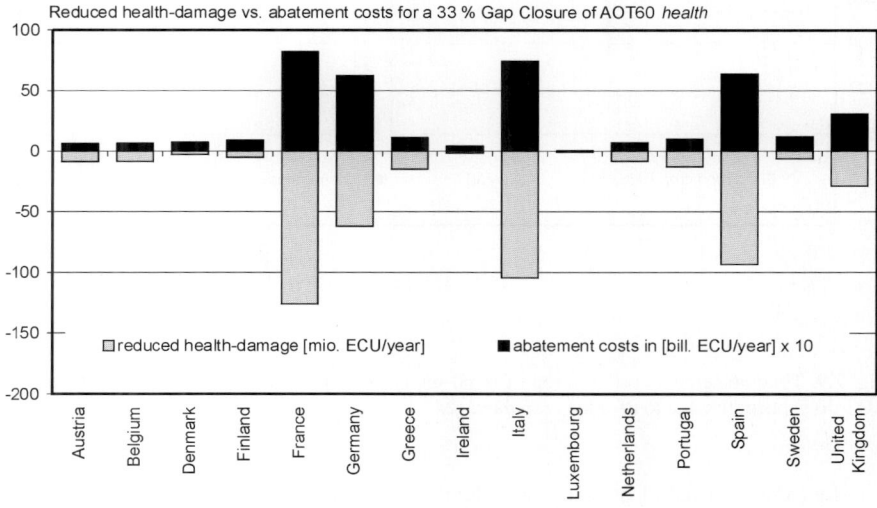

Fig. 7.8. Total abatement costs (on top of trend scenario 2010) to achieve 33 % Gap Closure of AOT 60 vs. benefits due to avoided health damages

Whilst having important implications for NO_x control in the high-NO_x countries, the acidification-ozone weighting has a more limited impact on the environmental improvements predicted by OMEGA. As shown in *Plates 14-16* the 15% and 33% GC scenarios give similar resulting AOT levels regardless of whether the acidification weighting is 0 or 0.5.

The total abatement costs depicted in *Fig. 7.7* indicate that the a Gap Closure of 10 % (1.4 bill. ECU) and 25 % (11 bill. ECU) is achievable at comparatively low total costs, while costs increase considerably for the 33 % GC case (39 bill. ECU). These costs of abatement are somewhat balanced by the monetary benefits from reduced damage costs from impacts of concentrations of ground-level ozone on health or crop yield, even though the monetarised damages are not as high as e.g. those caused by SO_2 or particulates (Friedrich and Krewitt 1997). *Figs. 7.8* and *7.9* show a abatement costs vs. reduced damage costs for health and crops (it has to be noted that the abatement costs are one order of a magnitude higher than the benefits). The calculation of health impacts includes accute effects, but – according to current, maybe insufficient knowledge – no relationship that relates increases in ozone concentrations to increases os chronic mortality (i.e. reduction of life expectancy).

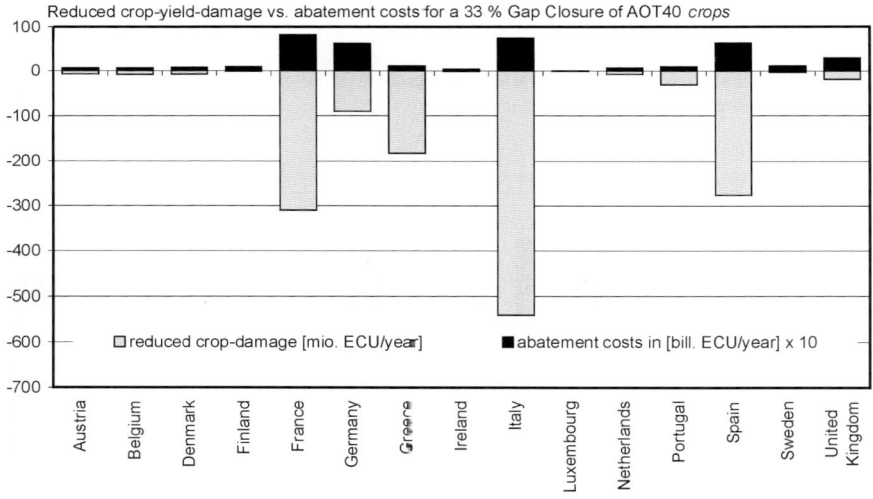

Fig. 7.9. Total abatement costs (on top of trend scenario 2010) to achieve 33 % Gap Closure of AOT 40 vs. benefits due to avoided crop damages

The fact that estimated benefits of reducing ozone are smaller than the associated costs does not necessarily imply that – under cost-benefit considerations – an implimentation of such a strategy is not useful. Firstly, reduction of ozone precursor emissions does not only reduce concentrations of ground-level ozone, but leads to the reduction of other harmful pollutants as well (e.g. fine particles PM 2.5, acidifying, eutrophying or cancerogeneous substances). These effects tend to have significantly higher health benefits than those calculated for ozone. This illustrates the problem of investigating one pollutant only, as such benefits could only be assessed in a consistent way by analysing a multi-effect abatement strategy. Secondly, as already hinted at, there may well be further damage caused by ozone, which is not accounted for in the state-of-the-art exposure-response relationships used in this study. As a consequence, the damage figures have not been used to conduct a cost-benefit analysis, but for the purpose of providing a relative basis for the development of a burden sharing rule according to the "victim pays" principle (cf. *Chap. 9*). The figures show, however, that the distribution of costs and benefits varies considerable, emphasising the need for burden sharing and a distributional analysis, which is *Chap. 9*.

In addition to the detailed investigation of scenarios, OMEGA was applied to calculate the emission reductions necessary for more stringent gap closures of AOT thresholds.

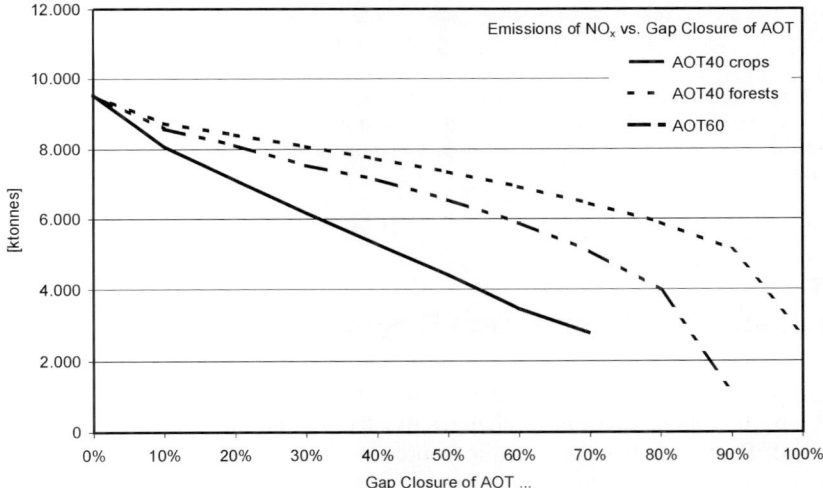

Fig. 7.10. Reductions of total EU15 NO_x emissions necessary to achieve specified gap closures of AOT40*crops*, AOT40*forests* and AOT60, starting from the Trend Scenario for 2010; for each 10 % gap closure step, emission reductions have been calculated with OMEGA

Fig. 7.10 indicates that a total EU15 NO_x emissions would have to be cut to about half of the trend scenario emissions for 2020 to achieve gap closures of about 50 % to 60 % of AOT40/60. As can be seen in *Fig. 7.11* as well, even a complete abatement of anthropogenic NMVOC emissions and considerable reduction of NO_x emissions would still result in exceedances of AOT40 for crops. AOT40*forests* seems to be less sensitive, probably because of the allowed exceedance of 10.000 ppm.h and the 6-month averaging period. For AOT60, a considerable gap closure of 80 % would be feasible by reducing both NO_x and NMVOC emissions by about 60 % (relative to the trend scenario).

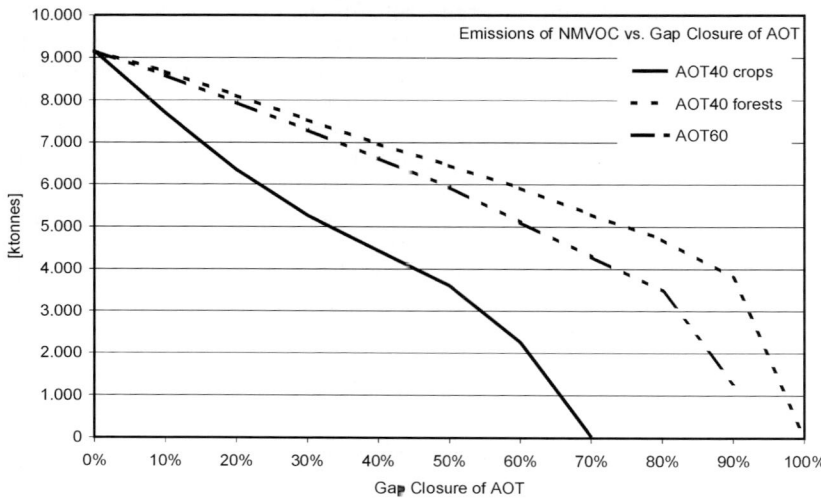

Fig. 7.11. Reductions of total (anthropogenic) EU15 NMVOC emissions necessary to achieve specified gap closures of $AOT40_{crops}$, $AOT40_{forests}$ and AOT60, starting from the Trend Scenario for 2010; for each 10 % gap closure step, emission reductions have been calculated with OMEGA

7.4
Evaluation of Optimisation Results

An important step in evaluating the potential usefulness of the iteration method is to ensure that its predictions of changes in excess pollution are similar to those of the EMEP models from which the source-receptor matrices were derived. For sulphur and nitrogen deposition this comparability is guaranteed, as the EMEP acid-deposition model is formulated as a linear system throughout.

Evaluation is clearly needed for ozone though, as the S-R relationships are known to be non-linear, especially in N.W. Europe. However, work with the EMEP photo-oxidant model has shown that for long-term statistics such as 6-monthly mean, or 3- or 6-monthly AOT values, there is a remarkable degree of linearity between ozone changes and emission reductions, especially with regard to VOC control (Simpson, 1991, 1992, Simpson and Malik, 1996).

Greater uncertainty arises in applying linear matrices to significant NO_x emission reductions (> 50%). In the iteration model this source of error is minimised by making the S-R matrices vary with NO_x emission level as discussed above, but some errors are inevitable in this treatment. Indeed, at the start of the iteration the source-receptor matrices used inside the model are identical to those produced by the EMEP model.

A preliminary evaluation of the iteration model, conducted generic cost curves, has been presented in Simpson and Eliassen (1997, 1999), and this showed

7.4 Evaluation of Optimisation Results

reasonable agreement for ΔAOT40 predictions. Here, a new evaluation of the model is presented, conducted for the more realistic conditions of the optimisation work. The values of AOT40 and AOT60 resulting from some of the 33 % gap-closure scenarios presented above, calculated by using the OMEGA model and the EMEP model, have been compared. Emissions were first determined using OMEGA to achieve the optimisation, then these were run through the EMEP model to produce a set of ´true´ AOT40 and AOT60 fields against which to compare the iteration model's approximation.

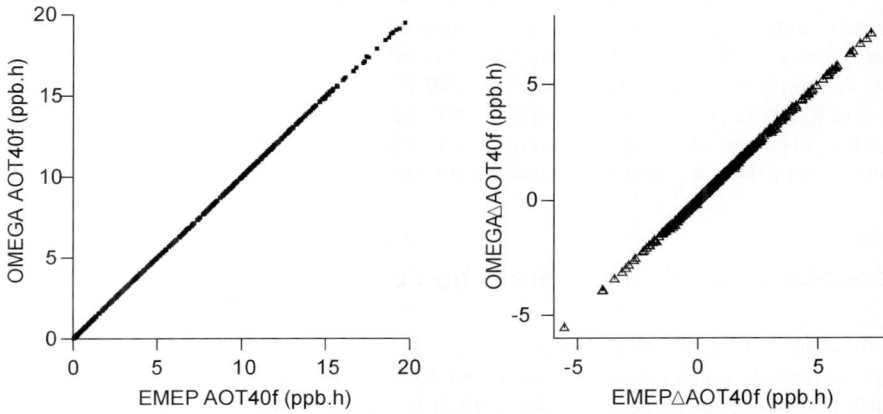

Fig. 7.12. Comparison of the absolute AOT40f values and the changes (reductions), ΔAOT40f, calculated by the OMEGA model and the full EMEP oxidant model. Each point represents one EMEP grids square. Results are taken from the 33% gap-closure scenario for AOT40f, for the ozone-only scenario ($w_{Ozone}=1.0$).

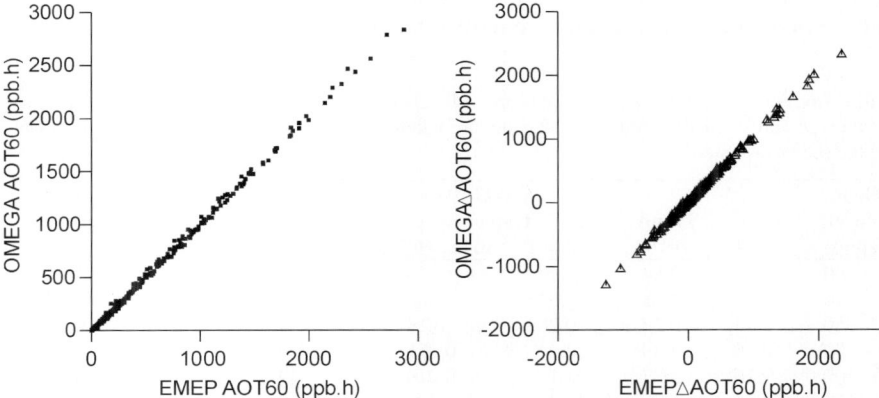

Fig. 7.13. Comparison of the absolute AOT60 values and the changes (reductions), ΔAOT60, calculated by the OMEGA model and the full EMEP oxidant model. Each point represents one EMEP grids square. Results are taken from the 33% gap-closure scenario for AOT60, for the joint acid-ozone scenario ($W_{Ozone}=0.5$).

As it is most important to evaluate if OMEGA correctly predicts relative changes in pollution fields correctly, both the AOT and ΔAOT values obtained from the OMEGA model have been plotted against the corresponding values obtained from the EMEP ozone model.

The results, shown in *Fig. 7.11* for AOT40 and *Fig. 7.12* for AOT60, show a remarkably good agreement. Both the absolute values of AOT and the changes are almost identical between the OMEGA and EMEP models. Indeed, the evaluation results obtained here are significantly better than those obtained by Simpson and Eliassen (1997, 1999).This is likely due to a better setup of the S-R matrices (tied closely to the range of emissions determined within this study), and more realistic cost-curves. It may also be due to the limited number of 15 countries involved in the calculations compared to the full EMEP-area calculations of Simpson and Eliassen. These results clearly indicate that the OMEGA model can be used with good confidence as an approximation to the full oxidant model, at least for the range of emission reductions considered for the 33 % gap-closure scenarios.

7.5
Modelled Exceedance Statistics of Ozone Thresholds

Although AOT-type statistics are very useful for devising and interpreting emission-control effects, it is also useful to consider the frequency with which particular ozone levels are exceeded with different scenarios. This frequency is not available from the OMEGA model's seasonal average statistics, but rather has to be obtained for a particular optimised scenario by re-running the full oxidant model. This re-run generates ozone concentrations every 6 hours. Although not easily comparable with the one-hourly or 8-hour average data specified in various ozone standards, here the frequency with which these EMEP model results exceed a range of ozone concentrations, from 60-120 ppb, is presented.

Table 7.6. Percentage of time for which ozone levels of 60-120 ppb are exceeded in the EU, as an average across all EU land-areas. Results are taken from the 5-year run of the EMEP model for the scenarios listed.

Ozone level (ppb)	Base 1990	Trend 2010	AOT40 scenarios			AOT60 scenarios	
			15 % Gap closure $w_{Ozone}= 1$	$w_{Ozone}= 0.5$	33 % GC $w_{Ozone}= 1$	33% Gap closure $w_{Ozone}= 1$	$w_{Ozone}= 0.5$
60	4.68	2.84	2.14	2.13	1.98	1.98	1.99
70	1.92	0.88	0.64	0.63	0.58	0.57	0.57
80	0.83	0.28	0.20	0.19	0.17	0.17	0.17
90	0.36	0.089	0.057	0.056	0.046	0.046	0.045
100	0.16	0.027	0.017	0.016	0.014	0.014	0.014
110	0.065	0.009	0.005	0.005	0.004	0.004	0.004
120	0.026	0.003	0.002	0.002	0.001	0.001	0.001

As shown in *Table 7.6*, exceedances of the various ozone levels are already reduced significantly from 1990 levels with the trend 2010 scenario. However, further significant reductions in exceedance frequency do occur for the optimisation scenarios, especially for the higher ozone levels. The AOT60 scenarios lead to greater reductions in these exceedance frequencies than the AOT40 scenarios. There is little difference between the ozone-only scenarios ($W_{Ozone}= 1.0$) and the joint ozone-acid scenarios ($W_{Ozone}= 0.5$).

7.6
Summary and Conclusions

Within this study, an iterative method, the OMEGA model, has been developed capable of dealing with the optimisation problem associated with the non-linear relationship between the NO_x and NMVOC emissions from many countries as precursors of ground-level ozone.. In addition to dealing with tropospheric (near-surface) ozone, the OMEGA model is able to include acidification into the optimisation. The model builds heavily upon available EMEP source-receptor relationships, with an interpolation procedure used to allow the ozone source-receptor relationships to vary with emission levels.

An important feature of the methodology is that interest in different environmental problems is specified clearly, through the use of weighting factors, e.g. W_{Acid} and W_{Ozone}. The methodology can be readily extended to consider other environmental problems. Further, although this work only considers weighting factors applied equally to all countries, it is also straightforward to allow some countries to express most interest in ozone whereas other could express most interest in acidification.

The model is designed to be transparent and flexible. The basic principles can be explained to both scientists and policy makers. This technique can be used to provide early guidance on various control strategy options, and allows rapid exploration of the many targets and measures possible with these approaches. It also demonstrates how even complex multi-dimensional environmental problems may be amenable to understandable solutions.

The OMEGA model has been used to investigate the relationships between the costs and environmental benefits for a range of scenarios, varying from low-ambition (10 % gap closure) to high-ambition and expensive 90% gap closure scenarios. The results show that for AOT60 gap closures of 10 – 25 % are achievable at comparatively low total costs, 1 to 10 bill. ECU, whereas costs increase considerably for a 33 % GC case (39 bill. ECU).

7.7
References

Alcamo J, Shaw R, Hordijk L, eds. (1990) The RAINS model of acidification. Science and Strategies in Europe. Dordrecht, The Netherlands: Kluwer Academic Publishers.
Amann M et al (1999), Final report on work done under EU acidification and ozone strategy. International Institute for Applied Systems Analysis (IIASA), Laxenburg, Austria
ApSimon H, Warren R, Wilson J. (1994) The abatement strategies assessment model - ASAM: applications to reductions of sulphur dioxide emissions across Europe. *Atmos Environ* 28:649--663.
Barrett K, Sandnes H. Acidifying transboundary air pollution. calculated field and budgets across Europe 1985-1995. (1996) In EMEP MSC-W Report 1/96, Transboundary air

pollution in Europe. Part 1: calculated fields and budgets of acid deposition and near surface ozone, Norwegian Meteorological Institute, eds. Barrett and Berge.

Friedrich R, Krewitt W (1997) Umwelt- und Gesundheitsschäden durch die Stromerzeugung – Externe Kosten von Stromerzeugungssystemen. Springer, Heidelberg

Grennfelt P, Hov O, Derwent R. (1994) Second generation abatement strategies for NOx, NH3, SO2 and VOCs. Ambio23:425--433.

Fuhrer, J., Skärby, L., and Ashmore, M.A. (1997) Critical levels for ozone effects on vegetation in Europe, *Environ.Poll.*, 97, pp 91-96.

Heyes C, Schöpp W, Amann M. (1996) A simplified model to predict long-term ozone concentrations in Europe: The core of an integrated assessment model. In: Workshop on the control of photochemical oxidants over Europe, 24-27 October 1995, St. Gallen, Switzerland. Federal Office of Environment, Forests and Landscape, Bern; pp 47–62.

Holland M, King K, Haworth A (1998) Economic Evaluation of the Control of Acidification and Ground-Level Ozone – Final Report for DGXI of the European Commission. AEA Technology, Culham

Olendrzynski K. (1997) Emissions. In EMEP MSC-W Report 1/97, Transboundary air pollution in Europe. Part 1: Emissions, dispersion and trends of acidifying and eutrophying agents, Norwegian Meteorological Institute, ed. E. Berge, pp 23–50.

Posch M, de Vries W, Hettelingh J. (1995) Critical loads of sulphur and nitrogen 1995. In, Calculation and mapping of critical thresholds in Europe, Status report 1995 Coordination Centre for Effects, RIVM,Bilthoven, The Netherlands, eds. M. Posch, P.A.M. de Smet, J.-P. Hettelingh and R.J. Downing.

Schöpp, W, Amann, M, Cofala, J, Heyes, C, and Klimont, Z,. (1999), Integrated assessment of European air pollution emission control strategies, *Environ. Model. & Software*, 14, pp 1–9.

SEI. (1991) An outline of the Stockholm Environment Institute's coordinated abatement strategy model, CASM. Stockholm Environment Institute, University of York, Heslington, York, UK.

Sillman, S., Logan, A., and Wofsy, S.C. (1990) The sensitivity of ozone to nitrogen oxides and hydrocarbons in regional ozone episodes, *J. Geophys. Res.*, 20, 1837-1851.

Simpson D. (1991) Long period modelling of photochemical oxidants in Europe. Some properties of targeted VOC emission reductions. Norwegian Meteorological Institute, EMEP MSC-W Note 1/91.

Simpson, D. (1992) Long period modelling of photochemical oxidants in Europe. A) Hydrocarbon reactivity and ozone formation in Europe. B) On the linearity of country-to-country ozone calculations in Europe, EMEP MSC-W Note 1/92, Norwegian Meteorological Institute, Oslo.

Simpson D, Eliassen A. (1997) Control strategies for ozone and acid deposition - an iterative approach. Norwegian Meteorological Institute, EMEP MSC-W Note 5/97.

Simpson D, Eliassen A. (1999, in press)Tackling multi-pollutant multi-effect problems - an iterative approach, The Sci. Tot. Environ.,.

Simpson, D. and Malik, S. (1996) Photochemical oxidant modelling and source-receptor relationships for ozone, In EMEP MSC-W Report 1/96, Norwgian Meteorological Institute, Oslo, Norway.

Simpson, D., Olendrzynski, K., Semb, A., Storen, E., and Unger, S. (1997) Photochemical oxidant modelling in Europe: multi-annual modelling and source-receptor relationships, Norwegian Meteorological Institute, EMEP MSC-W Report 3/97.

UN-ECE (1997) Health-effects of Ozone and Nitrogen Oxides in an Integrated Assessment of Air Pollution, Institute for Environment and Health, University of Leicester, Leicester, UK.

Acknowledgements

Thanks are due to Prof. Trond Iversen (Univ. Oslo) and Anton Eliassen (DNMI) for reviewing the methodology and valuable discussions. This work was financed by The Norwegian Ministry of the Environment and the EMEP project, as well as the DG XII project INFOS.

8 Tropospheric Ozone and Urban Air Quality

Nicolas Moussiopoulos, Peter Sahm, Paraskevi-Maria Tourlou, Theodoros Nitis, Abul Kalam Azad and Sofia Papalexiou

8.1 Introduction

Exposure to air pollution represents a serious problem in many cities all over the world. Despite the achievements in the reduction of *traditional* emissions (SO_2 and Particulate Matter), the majority of European cities still exceed air quality guidelines. Nowadays, photochemical air pollution causes most of the concern, and in this context attention is focused on ozone - as the most prominent photo-oxidant - and nitrogen dioxide (NO_2). According to the recent report *Air Pollution in Europe 1997* of the European Environment Agency, the EU ozone threshold value for the protection of human health (110 µg/m^3, 8 h average) is exceeded substantially (Jol and Kielland, 1997). Based on measurements at urban stations it was concluded that 80 % of the EU urban population is exposed to these exceedances.

At the regional scale, NO_x control was found to be effective in reducing ozone concentrations in the more remote areas of Europe (e.g. Derwent et al., 1994). As the ozone phenomenology in urban areas is different from the general pattern (in cities NO_x generally suppresses the local concentration of ozone), NO_x emission reductions may lead to ozone increases in the urban area whereas in the close vicinity of a conurbation the available VOCs are limiting ozone formation. In contrast, far away from the source area the local ozone production rate is limited by the availability of NO_x. *Fig 8.1* shows model results for the typical evolution of ozone downwind a city as predicted by the OFIS model (Moussiopoulos and Sahm, 1998). And as the spatial resolution of regional scale models ranges from 50 to 150 km, local ozone peaks and the specific conditions in the areas surrounding conurbations cannot be properly assessed at this stage, thus making it necessary to apply mesoscale models for the assessment of local scale effects.

Much attention has been paid to formulate conclusions regarding the relative effectiveness of various control measures, e.g., VOC vs. NO_x control in urban areas. Comparisons of historical trends in ozone and precursor concentrations with trends in precursor emission levels can provide insights into the relative effectiveness of control strategies. Chinkin et al. (1996) compared correlation coefficients for NO_x and VOC emissions vs. ambient ozone for metropolitan areas in the Northeast of the United States and found somewhat better correlation between VOC and ozone as compared to NO_x although the statistical significance of these comparisons was not explored. The correlation coefficients were low overall, indicating that annual emission changes fail to explain a significant percentage of the variability in meteorologically adjusted ozone trends.

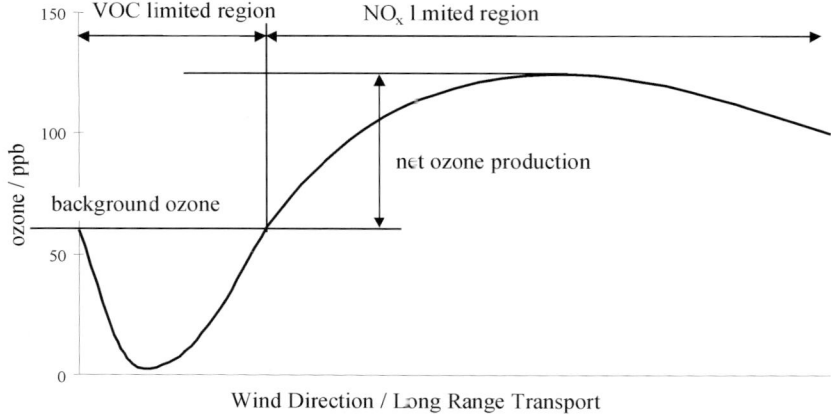

Fig. 8.1. Ozone concentration downwind a major source (e.g. a city).

8.2
Ambient Ozone Concentrations in Selected Mesoscale Areas (Athens, Stuttgart and Milan)

Effect-based research in Europe suggests using a long-term criterion as the no-damage and health-effect threshold (e.g. Kärenlampi and Skärby, 1998). As a consequence, integrated assessment should - besides short-term episodes - also address the long-term ozone exposure over a multi-month period. Air quality models have proven to be valuable tools for such investigations.

Thus, the EMEP model (Simpson 1993, 1995) was used to calculate ozone concentrations in Europe based on the reference year 1990 and the trend 2010 emission scenarios (see *Chap. 6*). In order to analyse local-to-regional scale effects, three mesoscale areas were further investigated with higher spatial resolution using the European Zooming Model (EZM). In particular, the ambient ozone concentrations over the Greater Athens area, the Greater Stuttgart area and the Milan metropolitan area were studied by the aid of the meteorological model MEMO and the photochemical model MARS. For the first two cases, simulations for three-month summer periods were performed, whereas, in the case of Milan, short-term episodic simulations were carried out. For each case, the ozone levels as well as the exceedances of guide values (see *Sect. 8.2.1*) were thoroughly examined considering the emission situation at two time horizons, namely in 1990 and in 2010. At this second case, three emission scenarios were adopted: The *business as usual* scenario and additional 50 % reductions of NO_x and VOC emissions on top of the *business as usual* scenario.

8.2.1
Emission Estimates

Athens Emission Inventory

The Athens emission inventory for the year 1990 was constructed on the basis of existing data on the emission situation in this year at a temporal and spatial resolution of 1 hour and 2 km, respectively. It comprises all available estimates of NO_x, CO, SO_2 and VOC emissions, the latter being allocated to selected groups of hydrocarbon compounds according to the source categories considered. In detail, the Athens emission inventory was compiled accounting for the following emission sources:

- Urban road traffic: The raw emission data related to urban traffic were prepared on a detailed Geographical Information System (GIS) and consisted of NO_x, CO and VOC emissions. Emission estimates for the year 1990 were based on an analytical method considering all available traffic data, calibrated by a "top-down" approach (Samaras et al., 1997).
- Industry: Industrial emissions were available in a gridded form and comprised NO_x, CO, SO_2 and VOC data. These emissions were adopted from the Auto Oil Study (European Commission, 1996). The calculation of the emissions was based on available data concerning fuel consumption (heavy oil, diesel and gas) of various industrial units and their location in each municipality of the Greater Athens area. Industrial emission sources were distinguished into area and point sources, the latter corresponding to five major plants.
- Airport: Emissions associated with the operation of the Athens airport were derived from information collected in the framework of the impact assessment study for the New Airport of Athens (Proyou et al., 1995).
- Space heating: Emissions estimations for space heating facilities were based on emission factors originally recommended by PERPA (Toll et al., 1995) and necessary activity data (i.e. the amount of fuel consumed for central heating in the area and its temporal variation) derived from available statistical information for 1990.
- Other sources: In addition to the above, extra-urban traffic and harbour emissions for NO_x, CO and VOC were also accounted for as in the Athens 2004 Air Quality Study (Moussiopoulos and Papagrigoriou, 1997). Considering the uncertainties regarding the spatial and temporal distribution and speciation of VOC emissions caused by fuel storage and other, mostly fugitive, VOC emissions, the total daily emissions for this source category were roughly estimated at 90 tons. It was assumed that 40% of these emissions are associated

with industrial activities, while the rest was distributed proportionally to the population density.

As a prerequisite for the synthesis of the Athens emission inventory, an emission inventory model was developed which compiles all available credible datasets concerning emission data for the present situation in the Greater Athens area. The main functions of this model are summarised as follows:

- Gridding process: All the aforementioned emission data were ultimately integrated in a 72×72 km^2 grid at a spatial resolution of 2 km. The discretisation of the urban traffic emissions was achieved using ARC/INFO (Environmental System Research Institute Inc., 1993). Emissions from the other source categories, i.e. emissions due to extra-urban traffic, industrial activities, harbour, airport and central heating were already available in a gridded, ready-to-use form.
- Temporal resolution: Since the diurnal variation of emissions has a strong influence on air pollution levels, the selection of an adequate temporal resolution is a crucial parameter with regard to the accuracy of an emission inventory. The Athens emission inventory contains data for each hour of the day considered as representative for the respective year.
- VOC speciation: The emission data collected comprised NO, NO_2, CO, SO_2 and VOC emissions. If the latter are to be applicable for any chemical mechanism embedded in a photochemical dispersion model, the emission inventory has to include accurate and detailed information on the VOC splitting. Consequently, the total VOC emissions were subdivided into 43 organic species. The distribution was accomplished by applying appropriate splitting factors - functions of the 43 species and the source category - to the VOC emission data for each source category. As far as road traffic, airport and harbour emissions are concerned, suggestions of Veldt (EMEP/CORINAIR, 1996) were adopted according to which different VOC speciation should be used for the various types of vehicle emissions. With regard to the speciation of industrial emissions the suggestions of Middleton et al. (1990) were followed.

Stuttgart Emission Inventory

Two emission inventories for the Greater Stuttgart area were generated by IER at a temporal resolution of 1 h and spatial resolutions of 5 and 2 km by applying the methodology described in brief as follows:

Emission data for the Greater Stuttgart area on a domain of 200×200 km^2 with Stuttgart in its centre were generated by IER and used as a basis for the intersection process with a GIS. Emissions from point, area and line sources were directly included in the gridded data set. Specific time curves were used for the sectors households and residential combustion (typical variations of heating patterns), industry (production indices and shift times), road transport (evaluation of information from roadside counters) power plants (utilisation profiles) and agriculture (typical variations of NH_3 evaporation). Finally, hourly emissions for each district were calculated using annual emissions, time-curve-shares (where applicable) and geographical emission shares, distinguishing point sources and area sources and giving specific emission values for NO_x, CO, NH_3, SO_2 and 32 different VOC species (Middleton et al., 1990).

The actual calculation has been conducted using a prototype version of the Emission Calculation Module (CAREAIR/ECM), which is currently being developed at IER (Friedrich et al., 1998).

Milan Emission Inventory

Simulations for the Milan metropolitan area were performed on the basis of a CORINAIR province-based emission inventory comprising emissions from road traffic and industry (large point sources were accounted for separately). Emissions coming from residential and commercial heating were not taken into account since a summer period was selected for the simulations. All necessary emission data for the Milan area were provided in the framework of the SATURN project. The emission information included CH_4, NH_3, NO, NO_2, CO, CO_2, SO_x and VOC emissions subdivided by activity sectors, being organised in two main parts, concerning respectively, area and main point sources. Emissions from area sources were available on a hourly basis and at spatial resolution of 4 km, whereas for the large industrial units (refineries and thermal power plants) information on the daily cycle profile was additionally provided. With the EZM being applied at a spatial resolution of 3 km, the area emissions had to be reallocated to the model resolution. For the speciation of the VOC emissions the methodology adopted for the generation of the Athens emission inventory was followed.

The biomass density and the emission rates adopted in all simulations for the calculation of biogenic VOC emissions were based upon the E-94 methodology. Corresponding data for isoprene were supplied by DNMI.

Emission Projections

The corresponding emission inventories for the *business as usual* trend scenario for 2010 were generated by applying attenuation emission factors to the 1990 base case emissions. These factors have been provided by IER, being consistent with the Europe-wide trend scenario described in detail in *Chap. 5*.

With regard to the Greater Athens area changes in both the amount and the spatial distribution of the emissions due to several infrastructure interventions that will take place in Athens by the year 2004, were additionally taken into account. The announced interventions for the Greater Athens area primarily involve the addition of new road axes, the operation of the new metro lines currently under construction and the reallocation of the Athens airport to the SE part of the simulation domain. With regard to the road traffic emissions, the penetration of future vehicle technologies in the Athenian car fleet –as imposed by the current legislation- was additionally accounted for.

For the Greater Stuttgart area, major changes in road traffic emissions due to increasing share of cleaner vehicles in the fleet have been accounted for, as well as significant decreases in emissions of stationary sources. The projection has been based upon research findings of previous projects investigating tropospheric ozone in the Upper Rhine valley, especially looking at transboundary transport of air pollutants from France, Switzerland and Germany (Obermeier et al., 1997) and detailed strategies to reduce NMVOC emissions in the federal state of Baden-Wuerttemberg (Berner et al., 1996).

Due to lack of information with regard to infrastructure changes to be completed within the specific time horizon, the Milan emission inventory for the *business as usual* scenario was constructed by applying the corresponding emission factors provided by IER to the 1990 base case emissions.

A detailed description of the methodology adopted for the projection of the 1990 emission estimates to the year 2010 may be found in *Sect. 5.1*.

8.2.2
Ambient Concentrations of Tropospheric Ozone in Athens

Athens Topography and Meteorology

As shown in *Fig. 8.2*, Athens is located in a basin of approximately 450 km². This basin is surrounded at three sides by fairly high mountains (Mt. Parnis, Mt. Pendeli, Mt. Hymettus and Mt. Aegaleon), while to the SW it is open to the sea. Industrial activities take place both in the Athens basin and in the neighbouring Thriasion plain (cf. solid areas in *Fig. 8.2*).

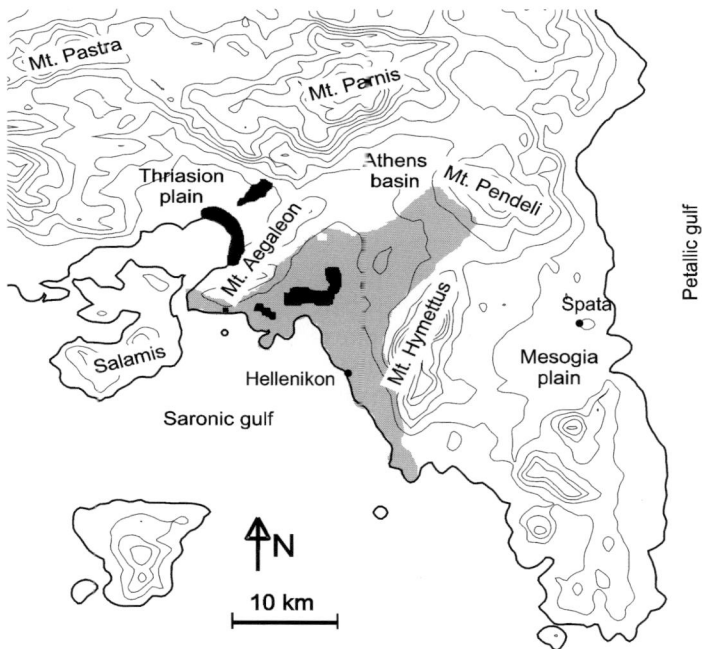

Fig. 8.2. Topography of the Greater Athens area. Altitude isopleths are contoured at 100 m. Residential areas are stippled, industrial areas in the Athens basin and in the Thriasion plain are solid.

Air pollution episodes occur in Athens repeatedly during all seasons of the year. Most of these episodes are associated with the development of sea breeze (Lalas et al., 1983), although the situation is probably even worse in case of stagnant conditions (i.e., a critical balance between synoptic and mesoscale circulations). Sea breeze cells develop not only in the Athens basin itself, but also in the Thriasion and Mesogia plains.The driving force for the intense sea breeze circulations in the Greater Athens area is the high insolation during summer days with anticyclonic weather conditions. The characteristics of the sea breeze in Athens are well known from previous observational work (Lalas et al. 1983; Lalas et al. 1987): Firstly the sea breeze tends to stratify the atmosphere above Athens thus trapping air pollutants at a relatively small height above ground. In addition, a recirculation of air pollutants takes place consisting of pollutant transport by the land breeze onto the sea and their re-advection back to the basin by the sea breeze; this results in an abrupt increase of the pollutant concentration levels in the Athens basin during the day. Apparently, in the case of chemically reacting pollutants (e.g. photochemical oxidant precursors) significant chemical transformations may occur in the course of this circulation.

On the other hand, radiational cooling frequently leads to very strong night-time inversions; apart from the associated high ground level pollutant concentrations, high stability causes large amounts of photochemical oxidants produced the previous day to persist at some height above the ground, where they cannot be diminished by surface NO_x emissions (Cvitaš et al. 1985).

Case Specification

The numerical grid used for the simulations in the case of the Greater Athens area covered an area of 72×72 km^2 at a horizontal grid resolution of 2 km, corresponding to 36 × 36 gridcells (*see Fig. 8.2*). Simulations were carried out for each day between the 20th of May and the 1st of September 1990.

Detailed orography data for the area of interest, needed for both meteorological and pollutant dispersion simulations, were derived from the GTOPO database, a global digital elevation model (DEM) with a horizontal grid spacing of 30 arc seconds (approximately 1 km) which was developed through an international collaborative effort led by staff at the U.S. Geological Survey's EROS Data Center (EDC). Land-use data originated from a combination of the CORINE Land Cover database, from which data on the continental part of the domain were drawn, and the Global Land Cover Characteristics (GLCC) Database (Belward, 1996) which covered the insular part of the study area. GLCC is a 1-km resolution global land cover characteristics data base generated by the U.S. Geological Survey (USGS), the University of Nebraska-Lincoln (UNL) and the European Commission's Joint Research Centre (JRC). CORINE Land Cover database includes 255 land cover species and uses the Lambert Azimuthal Equal Area geographical projection (optimised for Europe) with a horizontal resolution of 250 m. In the present application a subtotal which contains 7 land cover species with a horizontal resolution of 2 km was used.

Aspects related to the formation of photochemical oxidants were analysed with the multilayer model MARS, which was developed as a simplified version of the 3-D dispersion model MARS, in conjunction with the EMEP MSC-W chemical reaction mechanism. The model was applied on a 72×72 km² domain with a spatial resolution of 2 km. All simulations were performed for sequential eight-day time intervals covering the aforementioned period; the first 24 hours of each such interval corresponded to a start-up run aiming at a built-up of the pollutants in the air over the area considered. Five layers were considered in the vertical direction, the minimum spacing at ground level not exceeding 20 m. The upper layer was set to an altitude of 3000 m.

The necessary meteorological input was derived from results of MEMO. The meteorological simulations were driven by upper air data (wind speed and direction and temperature at ca. 700 m) derived from the results of the Norwegian Meteorological Institute's numerical weather prediction (NWP) model, which is described in detail in Grønas and Hellevik (1982), Grønas and Mitbo (1986) and Nordeng (1986). These data were available at a spatial resolution of 150 km and a temporal resolution of 6 hours. 6-hour average pollutant concentration values for the Greater Athens area with a spatial resolution of 150 km predicted by the EMEP model have been used as initial and lateral boundary conditions at inflow. Though defined as boundary layer averages, these values have been used for all model layers. The emission inventory upon which the photochemical simulations were based has been described in detail in *Sect. 8.2.2*.

Base Case Results

In *Fig. 8.3*, a set of four indicators for the assessment of the ozone exceedances (e.g. the accumulated exposure to ozone over the threshold value of 60 ppb, AOT60, (upper left), the number of days with the 8h average for ozone exceeding the limit value set in the frame of the ozone daughter directive of 120 µg/m³ (upper right), the number of hours with the 1h average exceeding 180 µg/m³ (lower left) and 240 µg/m³ (lower right)) as resulted by MARS simulations for the reference year 1990, are illustrated.

AOT60 values exceed 20 ppm×h in a restricted part of the domain, while in the urban area they are, in general, of the order of less than 2 ppm×h. Regarding the number of days of exceedance, they cover about 70 % of the period studied in a large part of the domain. The number of hours during which the limit of 180 µg/m³ is exceeded by the 1h average ozone concentrations reaches 160-180 in a limited part of the domain. The simulations reveal that highest concentrations exceeding 240 µg/m³ occur at the northeast and south east of Athens with a low but not negligible frequency.

As evidenced from the figure, the most seriously affected areas are those at the south west part of the Attica peninsula and, to a lesser extent, the northern suburbs of Athens. According to the regional scale meteorological synoptic information, the former area happens to be most of the time downwind of the city of Athens. The latter area is mainly affected when local circulation systems and particular sea breeze circulations prevail in the region. The urban area itself is exempt from serious exposures as the ozone is depleted in the presence of the intense urban NO_x emissions.

A more quantitative representation of these results is reported in *Table 8.1*. In particular, the maximum and domain averaged values of AOT60 and of the number of exceedance days as well as the 8h and 1h maximum and the three-month average ozone concentration values for the reference year are shown separately for the urban area, the suburban area and the whole of the study domain. Large exceedances of the ozone threshold values are predicted both at the suburban areas and as domain averages.

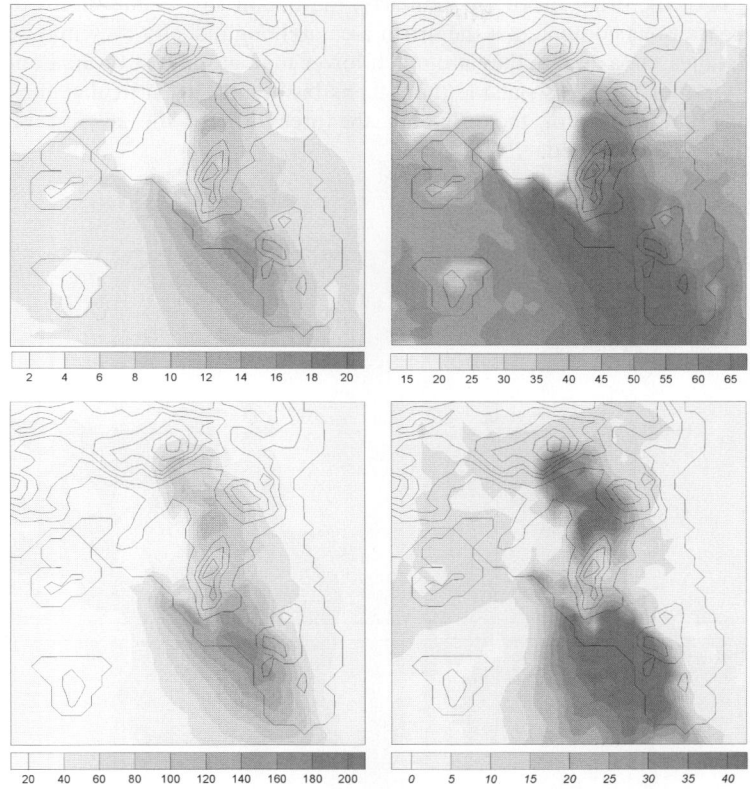

Fig. 8.3. Model results for the accumulated exposure to ozone over the threshold value of 60 ppb, AOT60 in ppm×h, (upper left), the number of days with the running 8h average for ozone exceeding 120 μg/m^3 (upper right) and the number of hours with 1h average ozone concentrations exceeding 180 (lower left) and 240 μg/m^3 (lower right) for the Greater Athens area according to the reference case. Altitude isopleths (black solid lines) are contoured at 150 m intervals.

Table 8.1. Model results for the Greater Athens area for the reference case 1990.

		Domain	Suburban	Urban
Days with 8h average > 120 μg/m^3	Domain maximum	81	79	38
	Domain average	41	36	9
AOT60 (ppm.h)	Domain maximum	15.9	13.9	6.3
	Domain average	5.1	6.2	0.7
Maximum 8h average (μg/m^3)			331	273	248
Maximum 1h average (μg/m$^3)$			571	570	476
3-month average (μg/m$^{3)}$			81	59	35

As already noted above, the urban area is less affected, although locally, the various indicators may also reach high values. Thus, while at specific locations of the urban area (mostly at the eastern part of the city) exceedances are predicted at approximately 40 % of the days of the studied period (106 days), on average, this percentage is of the order of 10 %. Similarly, the maximum 1h and 8h averaged ozone concentrations computed at the urban area, although locally they may reach quite high values, on average, they are much lower than at the rest of the domain. The densely populated areas outside the Athens basin and, in particular, those at the east and south east of the city, are the areas where the highest ozone concentrations are computed.

Impact of Emission Reductions

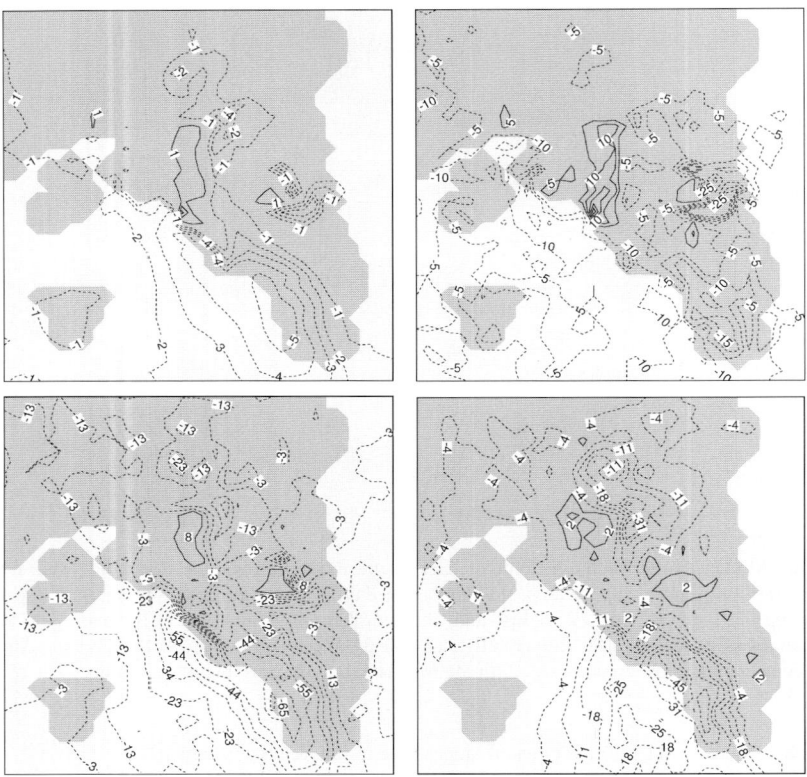

Fig. 8.4. Calculated differences between the "business as usual" 2010 scenario and the reference case in terms of AOT60 values (upper left), number of days of exceedance of the running 8h average of 120 µg/m^3 (upper right), number of hours with 1h average ozone exceeding 180 (lower left) and 240 µg/m^3 (lower right) for the Greater Athens area. Dashed curves denote decrease while continuous curves denote increase of the relevant values.

8.2 Ambient Ozone Concentrations in Selected Mesoscale Areas 131

Fig. 8.4 shows the difference with regard to 1990 in terms of AOT60 (upper left), number of days with the 8h average exceeding 180 μg/m^3 (upper right) and number of hours with the 1h mean exceeding 180 μg/m^3 (lower left) and 240 μg/m^3 (lower right) attained up to the time horizon of 2010, when assuming the *business as usual* emission scenario and taking into account all the infrastructure interventions planned to have been accomplished by 2010 in the city of Athens.

According to this scenario an overall reduction of 20 % in NO$_x$ emissions and of 50 % in VOC emissions is estimated. A quantitative assessment of the situation is given in *Table 8.2*, where the same indicators as in *Table 8.1* are reported.

Table 8.2. Model results for the Greater Athens area; *"business-as-usual"* scenario 2010.

		Domain	Suburban	Urban
Days with 8h average > 120 μg/m^3	Domain maximum	76	71	47
	Domain average	34	32	15
AOT60 (ppm.h)	Domain maximum	9.5	9.2	6.1
	Domain average	3.2	4.5	0.6
Maximum 8h average (μg/m^3)		257	257	230
Maximum 1h average (μg/m^3)		409	409	400
3-month average (μg/m^3)		78	61	45

From *Fig. 8.4*, it appears that the most important ozone level reductions are to be expected at the south east of the city of Athens, where the highest ozone burden was calculated for 1990. Thus, at these areas, both the maximum AOT60 value and the maximum number of days of exceedance are found to decrease by about 30 %. A considerable decrease (up to 80 %) in the frequency of the highest exceedances is also predicted at the northern suburbs of the city (see *Fig. 8.4*, lower right). At the same time, however, a doubling of the days of exceedance of 120 μg/m^3 by the 8h ozone mean is to be expected in the area close to the Hellenikon airport, where the human activities, and consequently the NO$_x$ emissions, will be drastically reduced as the airport will cease its service. The related increase of the AOT60 is much less pronounced and limited in the very vicinity of the airport indicating that the exceedances, although more frequent, will be of a smaller magnitude. A small increase of the number of exceedance days, of the order of 20 %, is also to be expected at the area of the new Spata airport, due to the increase of the overall emissions in the region. At the urban area itself, no important ozone level variations are to be anticipated.

The above findings are also reflected in *Table 8.2*. While for the suburban areas as well as over the whole of the domain both maximum and averaged indicator values are found to decrease, for the urban area an increase in the number of the days of exceedance is predicted. The AOT60, however, is expected to decrease indicating again that more frequent but less pronounced ozone threshold exceedances are to be expected in the city of Athens. This is also confirmed by the fact that both the 8h and 1h averaged maximum ozone concentrations at the urban area are lower in 2010 than in 1990, whereas the 3-month average increases.

In *Fig. 8.5*, the impact of a further 50 % reduction in NO$_x$ emissions on the local scale at the time horizon of 2010 is illustrated in terms of AOT60 (left) and number of days of exceedance (right). What is evidenced from the figure according to this scenario, is a strong increase of the average ozone levels, in particular, of the number of exceedance days and to a lesser extent of AOT60, above the urban part of the domain. Even with respect to the "business as usual" scenario (lower row of *Fig. 8.5*), the performance of the local NO_x emission reduction scenario is worse, leading in general to an increase of the AOT60 and the duration of the ozone exceedances.

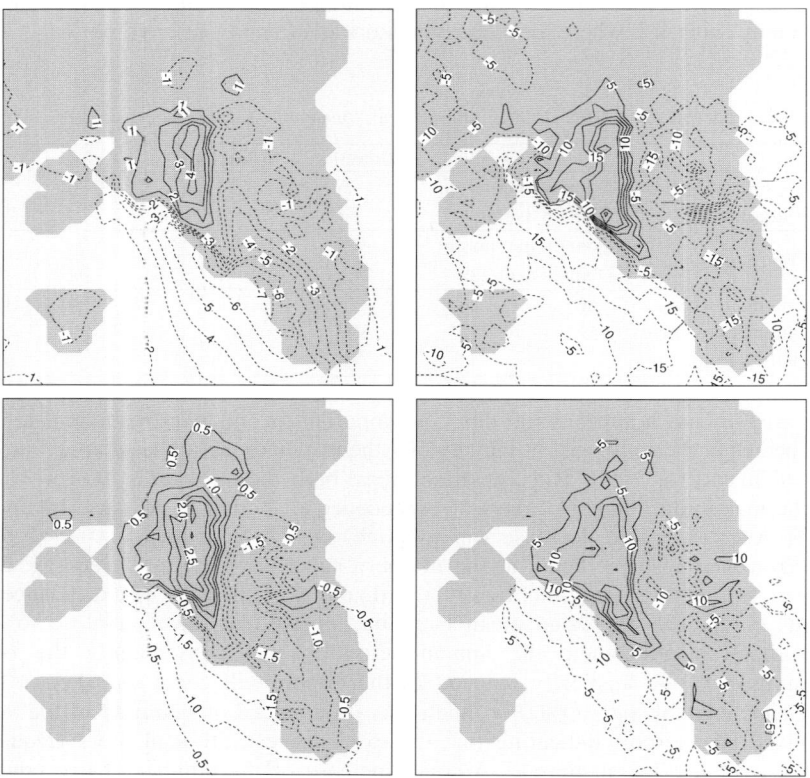

Fig. 8.5. *Upper row:* Calculated differences between the "business as usual" 2010 scenario enhanced by a 50% reduction in NO_x emissions and the reference case in terms of AOT60 values (left) and number of days of exceedance of the running 8h average of 120 µg/m³ (right) for the Greater Athens area. *Lower row:* Calculated differences between the additional 50% reduction in NO_x emissions on the local scale and the "business as usual" 2010 scenario in terms of AOT60 values (left) and number of days of exceedance of the running 8h average of 120 µg/m³ (right) for the Greater Athens area. Dashed curves denote decrease while continuous curves denote increase of the relevant values.

With the VOC emission levels remaining unchanged from the previous scenario, and the main ozone sink, i.e. NOx, being drastically reduced, both the number of days of exceedance and the AOT60 are almost doubled in a large part of the urban area. The same trend is apparent from the results presented in

8.2 Ambient Ozone Concentrations in Selected Mesoscale Areas

Table 8.3. The degree of exceedances is, however, lower as it results from the 8h and 1h maximum ozone concentrations reported in the same table.

Fig. 8.6 shows the differences in AOT60 (left) and the number of exceedance days of the ozone threshold (right) when alternatively applying an additional 50 % reduction in the emissions of VOC on the local scale at the time horizon of 2010. In contrast to the 50 % NO_x emission reduction of the previous case, a similar reduction in VOC emissions results in a further decline of the ozone burden at the urban site, with respect to the reference case, as compared to the decrease attained with the "business as usual" scenario (lower row of Fig. 8.6).

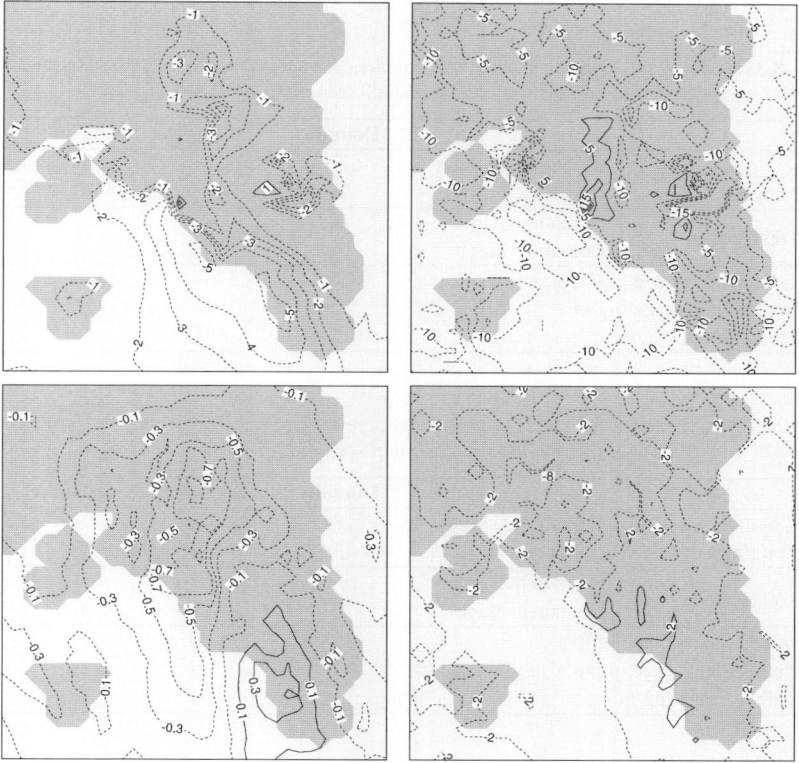

Fig. 8.6. *Upper row:* Calculated differences between the "business as usual" 2010 scenario enhanced by a 50% reduction in VOC emissions and the reference case in terms of AOT60 values (left) and number of days of exceedance of the running 8h average of 120 µg/m³ (right) for the Greater Athens area. *Lower row:* Calculated differences between the additional 50% reduction in VOC emissions on the local scale and the "business as usual" 2010 scenario in terms of AOT60 values (left) and number of days of exceedance of the running 8h average of 120 µg/m³ (right) for the Greater Athens area. Dashed curves denote decrease while continuous curves denote increase of the relevant values.

At the southwest of the Attica peninsula, however, an increase of both the AOT60 and the number of exceedance days, comparing to the "business as usual" scenario, is predicted. The overall gain with respect to the latter scenario is, therefore, marginal as is also evidenced by the indicators appearing in *Table 8.4*.

In conclusion, the ozone levels computed in the Athens area for the year 1990 are quite high and serious and durable exceedances of the threshold values are recorded. In the time horizon of 2010, considering the *business as usual* scenario and taking in parallel into account all the interventions that are either planned or are already under way in the city and will be completed by the year 2010, the situation is expected to ameliorate. These benefits are not counterbalanced by the rather insignificant ozone increase in the urban area itself (more frequent, but less pronounced, ozone thresholds exceedances).

Table 8.3. Model results for the Greater Athens area for the scenario with 50 % reduction in local NO_x emissions on top of the "business as usual" scenario for 2010.

		Domain	Suburban	Urban
Days with 8h average > 120 μg/m³	Domain maximum	76	75	68
	Domain average	36	40	30
AOT60 (ppm.h)	Domain maximum	9.4	9.4	8.7
	Domain average	3.2	5.1	1.3
Maximum 8h average (μg/m³)		276	276	260
Maximum 1h average (μg/m³)		426	393	426
3-month average (μg/m³)		81	69	61

Table 8.4. Model results for the Greater Athens area for the scenario with 50 % reduction in local VOC emissions on top of the *business as usual* scenario for 2010.

		Domain	Suburban	Urban
Days with 8h average > 120 μg/m³	Domain maximum	84	78	44
	Domain average	38	31	13
AOT60 (ppm.h)	Domain maximum	10.9	9.3	5.7
	Domain average	3.7	4.2	0.5
Maximum 8h average (μg/m³)		253	290	223
Maximum 1h average (μg/m³)		427	508	362
3-month average (μg/m³)		78	79	45

A further reduction of 50 % in NO_x emissions has an adverse effect on the ozone threshold exceedances in the urban area, while in the rest of the domain the situation remains more or less unchanged as compared to the "business as usual" scenario. On the contrary, an additional 50 % reduction in VOC emissions results in reductions of both the frequency and the degree of the exceedances in the urban area compared to the *business as usual* scenario.

8.2.3
Ambient Concentrations of Tropospheric Ozone in Stuttgart

Stuttgart is the capital of the German federal state of Baden-Wuerttemberg and the economic and industrial hub in the centre of Southern Germany. The city is located in a valley with steeply rising slopes embedded in the central Neckar region with its more than 2.5 million inhabitants. Traditional crafts and modern industry are the backbone of the economy. In and around Stuttgart are to be found the headquarters of automotive industries. Here, as everywhere else in Baden-Wuerttemberg, there is a highly organised network of small and medium-sized firms who supply parts and equipment to the big companies. Adjacent to the central Neckar industrial region are Karlsruhe with its oil refineries, Ludwigshafen with Europe's biggest chemical corporation, BASF, Mannheim and Heidelberg which make buses and printing machinery respectively, but also Freiburg and Ulm with their extensive service industries.

The temperate climate in the state of Baden-Wuerttemberg is characterised by its warm summers and mild winters. Ozone episodes with exceedances of 180 $\mu g/m^3$ frequently occur in the period between March and September. Moreover, the background ozone levels at remote areas of this state were found to increase by 5 $\mu g/m^3$ per year between 1984 and 1991 (Mayer and Schmidt, 1992). A map of the ozone limit exceedances derived from measurements of German Federal Environmental Agency and State monitoring network (e.g. UBA, 1999) shows that ozone exceedance frequencies are highest in Baden-Wuerttemberg and lowest in northern Germany. In general, an increase of the ozone burden can be observed from the north to the southwest. This increase cannot be sufficiently explained through meteorological conditions which favour the occurrence of photochemical smog alone. It is likely that long-range transport plays an important role. In contrast to the northern part of Germany, where in specific high pressure conditions rather 'clean' air masses are advected from the Baltic sea or Scandinavia, air masses moving in direction to southern Germany have passed over densely populated and highly industrialised areas with high ozone precursor emissions.

In 1990, a network of 60 stations was measuring ozone in the state of Baden-Wuerttemberg, from where data from 40 stations have been available during at least 80% of the daylight-hours between March and September. *Fig. 8.7* shows the days with the running 8h mean ozone concentration exceeding 120 $\mu g/m^3$ as derived from those measurements. The figure reveals that highest ozone burden prevail in the vicinity of Stuttgart, especially to the east of Stuttgart, but also in the southwestern part of the state. The figure furthermore exhibits that prevailing wind directions in the greater Stuttgart area are those from south-west and thus indicates that the high ozone burden in the east of Stuttgart is indeed associated with the urban plume, whereas the high ozone burden in the south-west of the domain is most likely related either to long range transport or high local emitters like the major motorway traversing.

Case Specification

At a resolution of 5 km, the selected model domain covered an area of 250×250 km² including parts of the Black Forest and the Swabian Alb in the south, the Rhine valley in the west and parts of the Odenwald in the northwest (*see Fig. 8.7*). Major motorways are traversing the domain in the west and north. Wind flow simulations were carried out with the MEMO model on the aforementioned 50×50 staggered grid for each day between the 20th of May and the 1st of September 1990. For that period, meteorological data (wind speed and direction and temperature at ca. 700 m) were derived, as in the case of Athens, from the Norwegian Meteorological Institute's numerical weather prediction (NWP) model at a spatial resolution of 150 km and a temporal resolution of 6 hours.

Detailed orography data for the area of interest, needed for both meteorological and pollutant dispersion simulations, were drawn from the GTOPO database with a horizontal resolution of 30 arc seconds (approximately 1 km). Land-use data originated from CORINE Land Cover database with a resolution of 1 km as well. In the present application, a subtotal of the database containing 17 land cover species was used.

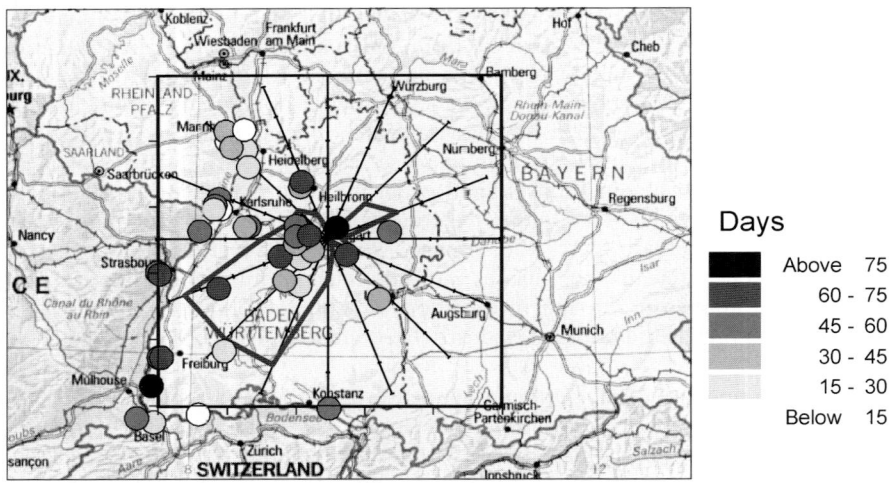

Fig. 8.7. Days with the running 8h mean ozone concentration exceeding 120 µg/m³ derived from measurements in the state of Baden-Wuerttemberg and wind direction statistics (at ca 700 m AGL) as derived from data supplied by DNMI between March and September 1990. The frame indicates the model domain for the EZM application.

Photochemical dispersion simulations were performed with the photochemical model MARS for the same time period using the EMEP MSC-W chemical reaction mechanism. As in the Athens case, all simulations were performed for sequential eight-day time intervals covering the aforementioned period; the first 24 hours of each such interval corresponded to a start-up run aiming at a built-up of the pollutants in the air over the area considered. Five layers were considered in

8.2 Ambient Ozone Concentrations in Selected Mesoscale Areas

the vertical direction, the minimum spacing at nd level not exceeding 20 m. The upper layer was set to an altitude of 3000 m.

The necessary meteorological input was derived from results of MEMO. 6-hour average pollutant concentration values for the Greater Stuttgart area predicted by the EMEP model have been used as initial and lateral boundary conditions at inflow. Though defined as boundary layer averages, these values have been used for all model layers. The emission inventory upon which the photochemical simulations were based has been described in detail in *Sect. 8.2.2*.

Base Case Results

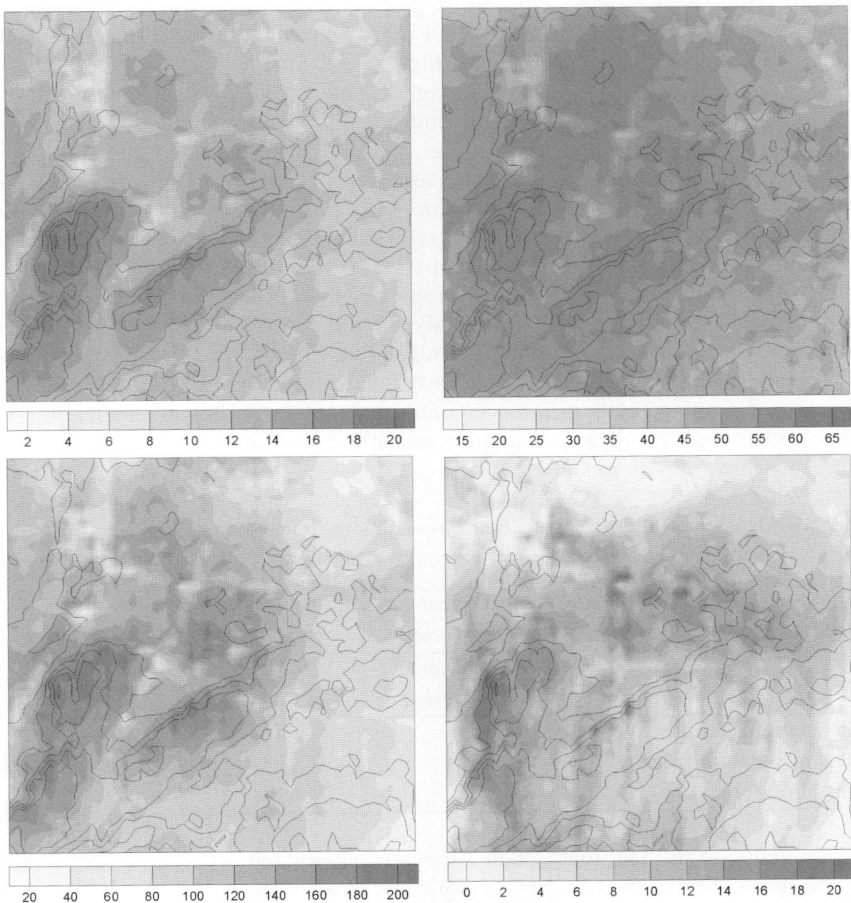

Fig. 8.8. Model results for the accumulated exposure to ozone over the threshold value of 60 ppb, AOT60 in ppm.h, (upper left), the number of days with the running 8h average for ozone exceeding 120 μg/m^3 (upper right) and the number of hours with 1h average ozone concentrations exceeding 180 (lower left) and 240 μg/m^3 (lower right) for the Greater Stuttgart area according to the reference case. Altitude isopleths (black solid lines) are contoured at 150 m intervals.

In *Fig. 8.8*, the accumulated exposure to ozone over the threshold value of 60 ppb, AOT60, (upper left), the number of days with the 8h average for ozone exceeding the air quality guideline of 120 µg/m^3 set in the EU Framework Directive (upper right), and the number of hours with the 1h mean exceeding 180 µg/m^3 (lower left) and 240 µg/m^3 (lower right) for the Greater Stuttgart area, as resulted by MARS simulations for the reference ye 1990, are shown.

As it is evident from the figure, a great part of the area suffers from high ozone levels, which exceed the threshold values for long periods. Thereby, during 60 % of the days of the study period, exceedances of the threshold of 120 µg/m^3 are predicted over the central part of the domain where the urban area is located, as well as over a large part of the region on the west and north west of the city of Stuttgart. This percentage becomes at ro part of the study area lower than 40 %. Highest values of the AOT60, of the order of 20 ppm×h, are predicted at the south west of the domain, areas where the highest absolute concentrations are also predicted (c.f. *Fig. 8.8* lower left and right). These high ozone levels are most likely related to the traffic emissions from the motorways in the south west of Stuttgart which are advected towards the centre of the domain by the south and southwest winds prevailing in the area, passing over parts of the Black Forest, an area with high biogenic emissions. In *Table 8.5*, a set of indicators aiming at the quantitative assessment of the ozone exceedances and exposure (in particular, maximum and domain averaged values of AOT60, number of exceedance days, 8h and 1h maximum and three-month average ozone concentration values) for the reference year 1990, are shown separately for the urban area, the suburban area and the whole of the study domain. Unlike the Athens case, where the ozone charge above the urban area was much lower than above the rest of the domain, in the case of Stuttgart the differences in terms of the indicators appearing in *Table 8.5* between the urban part and the rest of the domain are rather small.

Table 8.5. Model results for the Greater Stuttgart area for the reference case 1990.

		Domain	Suburban	Urban
Days with 8h average > 120 µg/m^3	Domain maximum	62	61	53
	Domain average	51	56	46
AOT60 (ppm.h)	Domain maximum	21.1	14.6	10.8
	Domain average	11.8	12.9	7.8
Maximum 8h average (g/m^3)		306	283	267
Maximum 1h average (g/m^3)		349	329	305
3-month average (g/m^3)		102	96	86

This is due to the fact that, while Athens is an isolated urban area surrounded by sea, in which the greatest part of pollutant emissions are accumulated in and around the city, in the case of Stuttgart the emissions are more uniformly dispersed over the whole of the domain. In addition, the regional ozone background concentrations in the area of Stuttgart are considerably higher than in the Athens area, as Stuttgart is located in a densely populated part of Central Europe. The maximum predicted ozone concentrations are, however, much lower than in the Athens case.

Impact of Emission Reductions

Fig. 8.9 shows the difference in terms of AOT60 (upper left), number of days of exceedance of the 120 µg/m^3 threshold (upper right) and number of hours with the 1h mean exceeding 180 µg/m^3 (lower left) and 240 µg/m^3 (lower right) attained up to the time horizon of 2010, if assuming the "business as usual" emission scenario in the Greater Stuttgart area. According to this scenario, an overall reduction of the order of 30 % in NO_x emissions and 40 % in VOC emissions is anticipated.

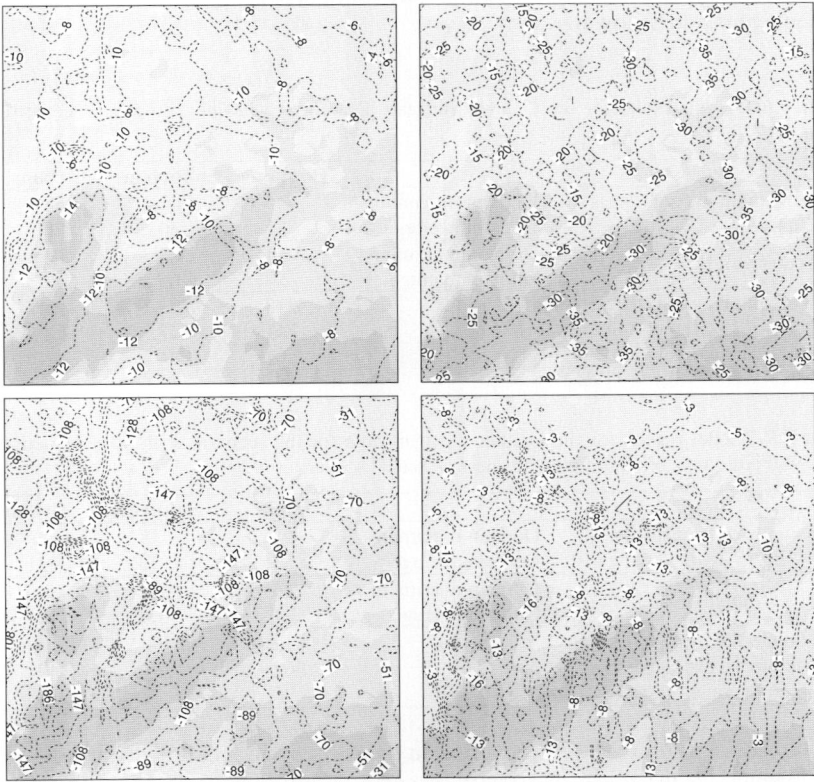

Fig. 8.9. Calculated differences between the *business as usual* 2010 scenario and the reference case in terms of AOT60 values (upper left), number of days of exceedance of the running 8h average of 120 µg/m^3 (upper right), number of hours with 1h average ozone exceeding 180 (lower left) and 240 µg/m^3 (lower right) for the Greater Stuttgart area. Dashed curves denote decrease while continuous curves denote increase of the relevant values.

As it results from the figure, the model predicts an important decrease up to 70 % in both the AOT60 and the number of days of exceedance of the 120 µg/m^3 threshold. Even bigger reductions are calculated in the case of the number of hours of exceedance of 180 µg/m^3 and 240 µg/m^3 by the 1h mean ozone concentration.

In contrast to the Athens case, there are practically no areas of increase of the indicators illustrated in the figures.

In *Table 8.6*, the same indicators as in *Table 8.5* are presented for the "business as usual" scenario of 2010. As it results from the figures in the table, there is a clear benefit with respect to all indicators, as both the degree and the duration of the exceedances are reduced. Thus, the average number of days of exceedance over the whole of the domain drops by 50 %, while, as far as the AOT60 is concerned, the gain is of the order of 80 %. The maximum concentrations decrease by about 30 % at the rest of the domain while at the urban areas this percentage is a little lower.

Table 8.6. Model results for the Greater Stuttgart area for the *business as usual* scenario for 2010.

		Domain	Suburban	Urban
Days with 8h average > 120 µg/m^3	Domain maximum	45	42	38
	Domain average	25	36	28
AOT60 (ppm.h)	Domain maximum	6.1	4.5	3.9
	Domain average	2.2	3.6	1.7
Maximum 8h average (µg/m^3)		213	197	195
Maximum 1h average (µg/m^3)		241	227	228
3-month average (µg/m^3)		87	84	77

Table 8.7. Model results for the Greater Stuttgart area for the scenario with 50% reduction in local NO$_x$ emissions on top of the "business as usual" scenario for 2010.

		Domain	Suburban	Urban
Days with 8h average > 120 µg/m^3	Domain maximum	48	38	39
	Domain average	23	32	31
AOT60 (ppm.h)	Domain maximum	5.6	3.7	3.9
	Domain average	2.1	3.0	2.1
Maximum 8h average (µg/m^3)		212	194	199
Maximum 1h average (µg/m^3)		363	341	308
3-month average (µg/m^3)		90	89	87

The impact of a further 50 % reduction in the local scale NO$_x$ emissions on top of the "business as usual" scenario with respect to the reference case is illustrated in *Fig. 8.10* in terms of AOT60 (upper left) and number of days of exceedance (upper right), while a similar comparison with respect to the "business as usual" scenario is shown in the lower row of the same figure. No important differences with respect to the latter scenario are evidenced. At the urban areas of Stuttgart, Karlsruhe, Mannheim, Heidelberg and Nurnberg, a small increase of both the AOT60 and the number of exceedance days is predicted. At the other parts of the domain, the ozone levels are further reduced. From *Table 8.7*, where the aforementioned indicators are presented for this scenario, it follows that the reduction of the local scale NO$_x$ emissions results in minor differences with respect to the "business as usual" scenario. At the urban areas both the duration and degree of the exceedances are slightly increased, while at the rest of the domain, however, the duration is slightly reduced as is evidenced by the lower AOT60 values.

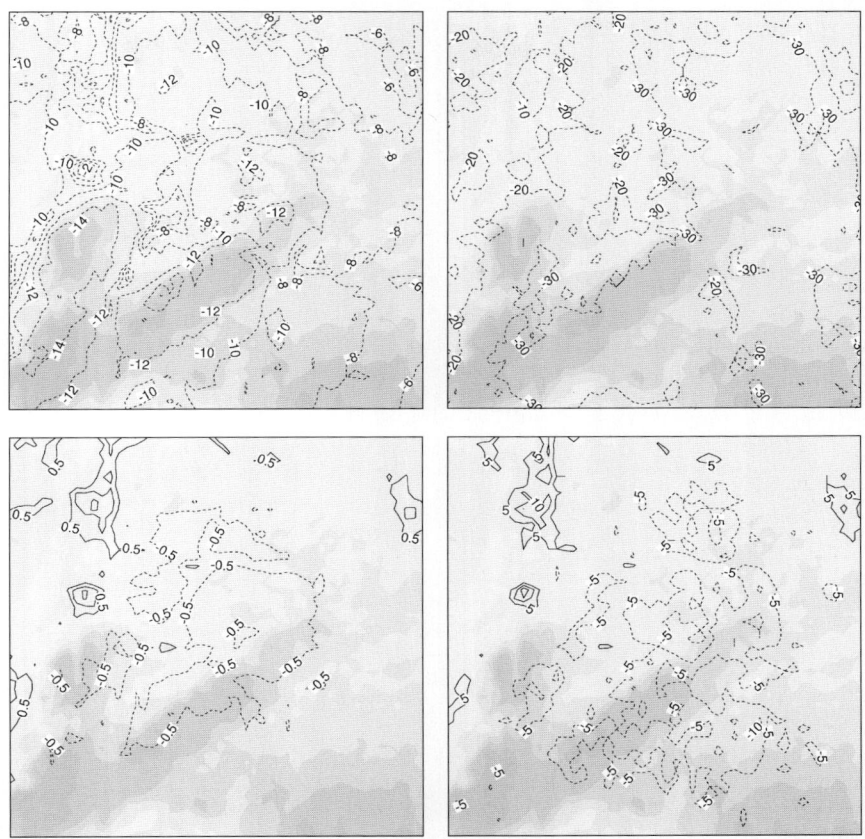

Fig. 8.10. *Upper row:* Calculated differences between the "business as usual" 2010 scenario enhanced by a 50% reduction in NO_x emissions and the reference case in terms of AOT60 values (left) and number of days of exceedance of the running 8h average of 120 µg/m^3 (right) for the Greater Stuttgart area. *Lower row:* Calculated differences between the additional 50% reduction in NO_x emissions on the local scale and the "business as usual" 2010 scenario in terms of AOT60 values (left) and number of days of exceedance of the running 8h average of 120 µg/m^3 (right) for the Greater Stuttgart area. Dashed curves denote decrease while continuous curves denote increase of the relevant values.

In contrast to the marginal impact attained with the 50 % reduction in local NO_x emissions, when reducing the VOC emissions at the local scale by 50 %, the result is a considerable decline of the ozone burden with respect to the "business as usual" scenario at the whole of the domain, as it follows from *Fig. 8.11* and *Table 8.8*. Both the intensity and the duration of the exceedances, as they are quantified by the indicators in *Table 8.8*, are further reduced by more than 20 % at all parts of the domain.

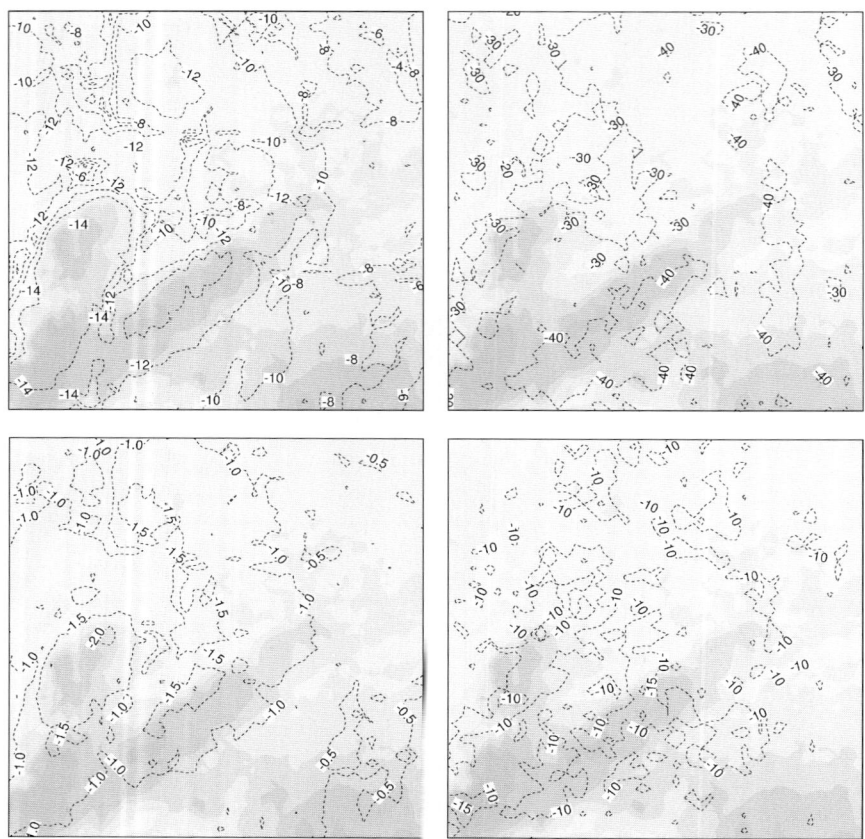

Fig. 8.11. *Upper row:* Calculated differences between the "business as usual" 2010 scenario enhanced by a 50% reduction in VOC emissions and the reference case in terms of AOT60 values (left) and number of days of exceedance of the running 8h average of 120 µg/m^3 (right) for the Greater Stuttgart area. *Lower row:* Calculated differences between the additional 50% reduction in VOC emissions on the local scale and the "business as usual" 2010 scenario in terms of AOT60 values (left) and number of days of exceedance of the running 8h average of 120 µg/m^3 (right) for the Greater Stuttgart area. Dashed curves denote decrease while continuous curves denote increase of the relevant values.

In conclusion, the Greater Stuttgart area suffers from serious and durable ozone exceedances. In the reference year 1990, exceedances are computed for a number of days which covers 60 % of the studied period. Unlike the Athens case, urban, suburban and rural parts of the domain do not experience very different ozone levels. For the time horizon of 2010 considering the "business as usual" scenario, quite important improvements of the air quality regarding ozone are predicted in the area.

Table 8.8. Model results for the Greater Stuttgart area for the scenario with 50% reduction in local VOC emissions on top of the "business as usual" scenario for 2010.

		Domain	Suburban	Urban
Days with 8h average > 120 µg/m^3	Domain maximum	40	32	28
	Domain average	17	26	18
AOT60 (ppm.h)	Domain maximum	4.3	2.6	2.4
	Domain average	1.2	2.1	1.0
Maximum 8h average (µg/m^3)		196	188	187
Maximum 1h average (µg/m^3)		227	212	217
3-month average (µg/m^3)		82	77	71

A 50 % reduction of the local scale NO_x emissions does not have a noticeable effect on the ozone levels. Absolute concentrations in particular seem to increase. On the contrary, a similar decrease of the local VOC emissions results in a further decrease of both the degree and the duration of the ozone exceedances by 20 – 30 %. Nevertheless, the ozone levels still remain high, exceeding the EU threshold during 30 – 40 % of the days of the investigated period.

8.2.4
Ambient Concentrations of Tropospheric Ozone in Milan

Milan Topography and Meteorology

With approximately 3 800 000 inhabitants the Milan metropolitan area is, in fact, the most industrialised and populated area of Northern Italy, frequently experiencing stagnant meteorological conditions associated with relatively high solar radiation. The city of Milan is located in a basin, the northern part of which is bounded by a portion of the Italian Alps. Consequently, the topography of the area is rather complex with an altitude increasing from 30 m to approximately 2870 m above sea level from south to north. The most inhabited and industrialised areas (e.g. the provinces of Varese, Como, Bergamo, Cremona, Pavia and Piacenza) are located around the city of Milan forming the so-called Milan metropolitan area.

As a physical barrier to the north of the area, the Alps protect the region from the intense air mass flows coming either from northern Europe or from the Mediterranean Sea. During the summer period, weak mountain-valley circulations prevail. These quite typical breeze circulations tend to come from the south during daytime (valley breeze) and from the north during night hours (mountain breeze) and may well be attributed to the temperature discontinuities between mountains and the lowland plains in the central-southern part of the domain. Calms or weak winds occur in approximately 80% of the days in the cold season of the year and in almost 40% of the summer days (Lavecchia et al 1996).

Case Specification

Wind field simulations were carried out using the MEMO model on a 120×180 km^2 domain around Milan at a spatial resolution of 3 km. 25 non-equidistantly distributed layers were distinguished in the vertical direction. The simulated period started at 2 LST on the 1st of July 1990 and ended at 22 LST on the 3rd of July 1990. This period was characterised by low winds and pollutant concentrations next to or exceeding both the ozone and nitrogen dioxide air quality standards at several stations of the permanent air quality monitoring network of Milan.

The simulations were driven by upper air data derived by soundings in the area of Milan. Detailed orography data for the area of interest, needed for both meteorological and pollutant dispersion simulations, came from GTOPO database. Land use data originated from GLCC database.

The simulations of pollutant dispersion and transformation were performed on a 96 × 96 km^2 domain around the Milan agglomeration by the aid of MARS model with the same horizontal resolution as the meteorological simulations. All relevant meteorological data (3-dimensional wind fields, potential temperature, turbulent kinetic energy, surface roughness, Monin-Obukhov length and friction velocity) were derived for pre-defined times from the simulation results of MEMO. Between these individual input times a temporal interpolation was performed. Photolysis rates were computed as functions of the solar zenith angle.

The dispersion simulations were performed using the EMEP chemical reaction mechanism. Initial concentration fields were produced during the first day of the simulations (the 1st of July 1990) which served as a start up run. Background concentrations consistent with results of the large-scale model EMEP for the area of Milan were used as initial and lateral boundary conditions.

Results

Model results for the wind field and the ozone concentrations at ground level at 12 and 16LST on the 2nd of July 1991 are presented in *Fig. 8.12*. Highest concentrations of the order of 160 μg/m^3 are predicted at 16 LST at the areas to the north east of the city of Milan as the urban plum is advected by the southwest winds that prevail towards this part of the domain. In *Table 8.9*, model results for the maximum and average ozone concentrations are reported. The threshold of 120 μg/m^3 is exceeded by the 8h mean ozone concentration at all parts of the domain although the exceedance at the urban area is marginal. In all cases, the ozone levels in the urban area are lower than in the rest of the domain.

8.2 Ambient Ozone Concentrations in Selected Mesoscale Areas

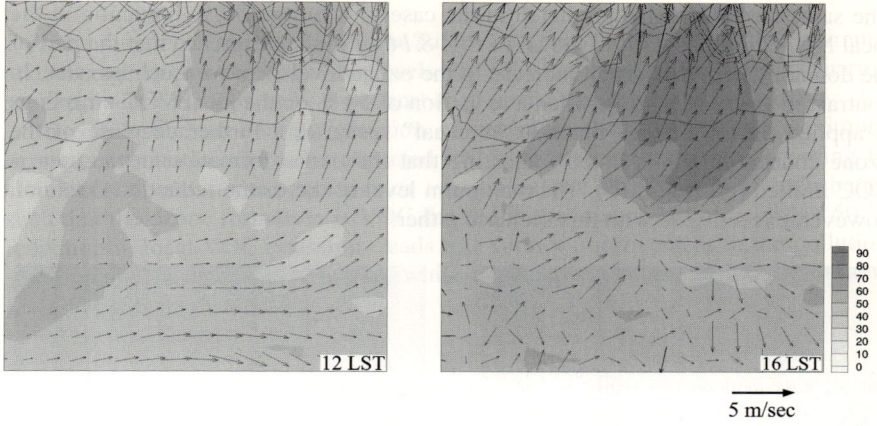

Fig. 8.12. Model results for the wind field and ozone concentrations (ppb) at a height of approximately 10 m above ground level at 12 and 16 LST on July 2, 1991 for the Milan metropolitan area according to the reference case. Altitude isopleths (black solid lines) are contoured at 150 m intervals.

Model predictions for the impact of the emission reductions on the ozone levels according to the *business as usual* scenario are shown in *Fig. 8.13*. Maximum reductions of the ozone levels of the order of 10 % are predicted at 16 LST at the areas where the maximum concentrations appear. At 12 LST when the NO_x emissions in the city area are highest, an increase in the ozone levels is predicted at the urban part of the domain.

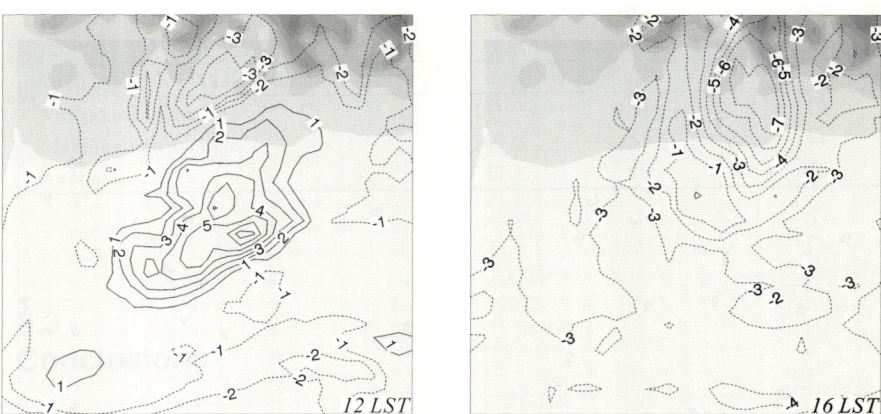

Fig. 8.13. Calculated differences of ozone concentrations (ppb) between the "business as usual" 2010 scenario and the reference case for the Milan metropolitan area. Dashed curves denote decrease whereas continuous curves denote increase of the predicted ozone values.

1990 and three future emission scenarios corresponding to the year 2010 and specifically, the "business as usual" scenario and additional 50 % reductions of NO_x and VOC emissions on top of the *business as usual* scenario.

The ozone levels computed in all three areas of concern exceed by far the threshold of 120 µg/m^3 for the 8h mean concentration that is set in the recently released EU ozone directive, for a number of days that can cover up to 80 % of the period studied in the case of Athens and 60 % in the case of Stuttgart. The urban areas in all three cases exhibit much lower ozone levels due to the depletion of ozone in the presence of the intense urban NO_x emissions. The most serious and durable exceedances are predicted for the Greater Athens area. On the other hand, the extent of the affected areas is much higher in the case of Stuttgart than in the case of Athens. This is justified by the fact that Athens is an isolated urban area sea, with the pollutant emissions accumulated in and around the city, while Stuttgart is located in a densely populated part of Central Europe where the emissions are more uniformly distributed and the regional ozone background concentrations are considerably higher. For this reason, the differences in ozone concentration between the urban areas and the rest of the domain are more pronounced in the case of Athens than in the case of Stuttgart. Milan resembles more to Stuttgart than to Athens.

In the time horizon of 2010, considering the "business as usual" scenario leads to important improvements of the air quality regarding ozone in all three areas. The benefit is larger in the case of Stuttgart where all ozone indicators are expected to decrease. A reduction of the order of 80 % of AOT60 is predicted at the urban site of the Greater Stuttgart area. For the Greater Athens area, at the urban cells ozone levels are even expected to increase slightly, while in the rest of the domain these levels are in generally reduced (40 % reduction of AOT60 for the rest of the domain). For the Milan metropolitan area an overall reduction of the order of 10 % in the maximum ozone concentrations is predicted.

A further reduction of 50 % in the local NO_x emissions has an adverse effect on the ozone levels at the urban areas, while in the other parts of the domains it does not have a noticeable effect as compared to the "business as usual" scenario in all three study areas. On the contrary, an additional 50 % reduction in VOC emissions results in reductions of both the frequency and the degree of exceedances in the three areas compared to the "business as usual" scenario. The results indicate that it is important to combine emission reduction measures at both the regional and the local scale, even for an isolated area like Athens.

Nevertheless, the ozone levels at the urban areas remain high with all three future emission scenarios. Even with the further 50 % reduction in VOC emissions, which appears to be the most efficient scenario for the ozone abatement, during the 103-day period studied, the 8h mean ozone concentrations still exceed the threshold of 120 µg/m^3 during 13 days at the urban sites of the Greater Athens area and 18 days at the urban sites of the Greater Stuttgart area. Even longer exceedances are predicted at non-urban sites of the two domains. More efficient emission reductions are, thus, necessary in order to achieve the targets defined in the Framework Directive of the EU with regard to ozone.

8.4 References

Belward A S ed. (1996) The IGBP-DIS global 1 km land cover data set (DISCover)-proposal and implementation plans. IGBP-DIS Working Paper No. 13, Toulouse, France

Berner P, Obermeier A, Friedrich R et al. (1996) Strategien zur Minderung der VOC-Emissionen ausgewaehlter Emittentengruppen in Baden-Wuerttemberg. FZKA-PEF 147, Oktober 1996, edited by: Forschungszentrum Karlsruhe GmbH, Karlsruhe 1996

Chinkin L, Reiss R, Eisenger D, Dye T and Jones C (1996) Ozone Exceedances Data Analysis: Representativeness of 1995. Final Report, Volume III: Summaries of Individual AQAWG Analyses OTAG Air Quality Analysis Workgroup, 1997

Cope M., Carnovale F., and Cook B (1992) The impact of emissions from gas-fired turbines for cogeneration on ambient air quality in Melbourne. in: Proceedings, 11th International Conference, Clean Air Society of Australia and New Zealand, Brisbane, 1, pp 328-37

Cvitaš T, Güsten H, Heinrich G, Klasinc L, Lalas D.P. and Petrakis M. (1985) Characteristics of air pollution during the summer in Athens, Greece. in: Staub - Reinh. Luft 45, pp 297-301

Derwent R G and Davies T J (1994) Modelling the impact of NO_x or hydrocarbon control on photochemical ozone in Europe, in: *Atmos. Environ.* **28**, pp 2039-2052

EMEP/CORINAIR (1996) Atmospheric Emission Inventory Guidebook. McInnes G (cd), First Edition, European Environment Agency, Copenhagen

Environmental System Research Institute, Inc. (1993) ARC/INFO Version 7.0.3, 380 New York Street, Redlands, CA 92373, USA

European Commission (1996) The European Auto Oil Programme, Report by the Directorate Generals for Industry, Energy and Environment, Civil Protection and Nuclear Safety

Friedrich R, Heidegger A and Kudermann F (1998) Development of an Emission Calculation Module as a Part of a Model Network for Regional Atmospheric Modelling. EUROTRAC Symposium 1998 Proceedings, *to be published*.

Grønas S and Hellevik O (1982) A limited area prediction model at the Norwegian Meteorological Institute. Norwegian Meteorological Institute Technical Report No. 61.

Grønas S and Mitbø K H (1986) Four dimensional data assimilation at the Norwegian Meteorological Institute. Norwegian Meteorological Institute Technical Report No. 66.

Grösslinger E, Radunsky K and Ritter M (1996) CORINAIR 1990 Summary Report 1. EEA Topic Report 7/1996. European Environment Agency, Copenhagen, Denmark

Jol A and Kielland G eds (1997) Air Pollution in Europe 1997. EEA, Copenhagen

Kärenlampi L and Skärb L eds (1998) Critical levels for Ozone in Europe: Testing and Finalizing the Concepts. UNECE Workshop Report, University of Kuopio, Department of Ecology and Environmental Science, Finland

Lalas D P, Asimakopoulos D N, Deligiorgi D G and Helmis C G (1983) Sea breeze circulation and photochemical pollution in Athens, Greece. in :Atmospheric Environment. 17, pp 1621-1632

Lalas D P, Tombrou-Tsella M, Petrakis M, Asimakopoulos D N and Helmis C G (1987) An experimental study of the horizontal and vertical distribution of ozone over Athens. in: Atmoperic Environment 21, pp 2681-2693

Lavecchia C, Angelino E, Bedogni M, Bravetti E, Gualdi R, Lanzani G, Musitelli A and Valentini M (1996) The ozone patterns in the aerological basin of Milan (Italy). *in: Environmental Software* 11, No 1-3, pp 73-80

Mayer H and Schmidt J (1992) Trendanalyse von Immissionsdaten in Baden-Württemberg. LFU Karlsruhe

Middleton P, Stockwell W R and Carter W P (1990) Aggregation and analysis of volatile organic compounds emissions for regional modelling. in: Atmospheric Environment 24A, pp 1107-1133

Moussiopoulos N and Papagrigoriou S, eds (1997) Athens 2004 Air Quality. Proceedings of the International Scientific Workshop "Athens 2004 Air Quality Study", Zappion, 18/19 Feb. 1997 (electronic version cf. www.envirocomp.org/html/publish/CDROM/Athens/flyer.pdf)

Moussiopoulos N and Sahm P (1998) The OFIS model: An efficient tool for assessing ozone exposure and evaluating air pollution abatement strategies. submitted to International Journal of Environment and Pollution

Nordeng T E (1986) Parameterization of physical processes in a three-dimensional numerical weather prediction model, Norwegian Meteorological Institute Technical Report No. 65

Obermeier A, Friedrich R, John C et al (1997) Czonproblematik im suedlichen Oberrheingraben: Emissionen, Minderungsszenarien und Immissionen. FZKA-PEF 162, edited by: Forschungszentrum Karlsruhe GmbH, Karlsruhe

Proyou A, Moussiopoulos N and Karatzas K (1995) Air quality impact caused by the transfer of the Athens airport from Hellenikon to Spata. in: Air Pollution III (Power H, Moussiopoulos N and Brebbia C A, eds), Computational Mechanics Publications, Southampton, Vol. 3, 131-139

Samaras Z, Andrias A, Zachariadis T and Aslanoglou M (1997) Forecast of road traffic emission for the Greater Athens area. Proceedings of the International Scientific Workshop "Athens 2004 Air Quality Study", Zappion, 18/19 Feb. 1997, pp 39-50

Simpson, D. , 1993, Photochemical model calculations over Europe for two extended summer periods: 1985 and 1989. Model results and comparisons with observations, Atmos. Environ., 27A, No. 6, 921-943.

Simpson, D. , 1995, Biogenic emissions in Europe 2: Implications for ozone control strategies, J. Geophys. Res., 100, No. D11, 22891-22906.

Toll I, Moussiopoulos N, Sahm P, Samaras Z, Zachariadis T and Aslanoglou M (1995) A new emission inventory for Athens and its use to improve air quality predictions. in: Air Pollution III, Vol. 2 (Power H, Moussiopoulos N and Brebbia C A, eds), pp 309-319

UBA (1999) http://www.umweltbundesamt.de/uba-info-daten-e/daten-e/k-o3180.htm, Umweltbundesamt, Berlin

9 Efficiency, Equity and Burden–Sharing

Roger Salmons

9.1
Introduction

This chapter focuses on the three inter-related issues of efficiency, equity and burden-sharing. There is a growing recognition that for international environmental problems[1] – where the achievement of an environmental goal requires the co-ordination of actions among separate sovereign states – it is important that any policy initiative[2] should not only be efficient (i.e. that the goal is achieved at least cost) but also that it must be equitable (i.e. that the burdens associated with achieving the goal are distributed fairly). As Rose (1992) observes in relation to global warming, "[While] equity considerations are usually accorded a secondary role in most economic policy-making, in the case of global warming, there are reasons why they may be paramount". This view is reinforced by Banuri *et al* (1996) who postulate that, in order to gain widespread participation, international agreements on reducing pollution must be perceived as being equitable – particularly among regions and countries.

Of course, this view has been developed in the context of international agreements where there is no supranational institution that can enforce the policy initiative without the willing co-operation of all the countries involved (e.g. the Kyoto Protocol on greenhouse gases, or the two UN/ECE Sulphur Protocols[3]). In the case of European Union policy on air pollution the situation is somewhat different. It is possible for a Directive to be adopted by qualified majority voting

---

[1] The term encompasses both global environmental problems such as climate change, and regional transboundary problems such as acid rain and tropospheric ozone pollution. The focus of this chapter is on the latter.
[2] The discussion will focus on international agreements where emissions reduction targets (i.e. abatement targets) are set for each country. However, the principles are equally applicable to other policy initiatives such as common technology standards, or international environmental taxes.
[3] Indeed, Atkinson (1998) suggests that the reason that the First Sulphur Protocol was only signed by a sub-set of the UN/ECE countries was that it was considered by the non-signatories to be unfair.

and imposed on all member states. However, in practice it is unlikely that a proposal would receive the necessary level of support if it were perceived as being inequitable in terms of the financial burdens that it imposed on the individual member states.

While equity issues have greater prominence for international environmental problems, this does not diminish the importance of efficiency. An agreement that imposes excessive costs is likely to face broad opposition, even if the distribution of these costs is perceived as fair. Unfortunately, the twin objectives of efficiency and equity will usually lead to different conclusions regarding the distribution of abatement effort between countries. An equitable allocation of abatement efforts (e.g. one that requires higher levels of abatement from richer countries) is unlikely to minimize the total cost of achieving the desired environmental objectives. Conversely, if abatement efforts are allocated so as to minimize total cost, then the resultant distribution of burdens is unlikely to be equitable. Burden-sharing mechanisms allow the reconciliation of these conflicting objectives. By introducing financial transfers between countries, they break the link between "who undertakes the abatement" and "who pays for the abatement".

The chapter is divided into two parts. The first deals with the issue of efficiency. It shows how a new form of permit trading – *iterative zonal permit trading* – can be used to reduce the cost of achieving desired reductions in tropospheric ozone concentrations across Europe. Equity and burden-sharing are considered in the second part of the chapter. In particular, the "efficient" cost allocation that has been derived for the 33 % Gap Closure scenario for AOT 60 (see *Chap. 7*) is compared with a number of alternative "fair" distributions, and the resultant burden-sharing transfers calculated.

9.2
Efficiency

This first part of the chapter considers how permit trading could be used to improve the cost efficiency of a European wide programme to reduce tropospheric ozone concentrations. It starts with a brief overview of the salient characteristics of pollution problems, and how these affect the design and practicality of trading systems. This is followed by a review of the ozone control problem, and how this can be simplified to a "linearized" local control problem that is amenable to permit trading. In the third section, a new form of permit trading – *iterative zonal permit trading* – is described, and its properties discussed. Finally, an illustrative example is used to show how iterative zonal permit trading might be applied to the case of tropospheric ozone.

9.2.1
Permit Trading and Pollution Characteristics

Pollution problems can be classified using many different taxonomies. However, when considering the potential for using permit trading systems to implement policy objectives such as air quality standards, there are three attributes of the pollution problem which are crucial to the design and viability of the alternative approaches. These are:

- the *uniformity* of emissions: does a reallocation of total emissions between sources and / or precursors affect the level of pollution at any receptor point?
- the *linearity* of emissions: does the impact of a change in emissions of a particular precursor at a particular source depend on its level?
- the *synergy* of emissions: does the impact of a change in emissions of a particular precursor at a particular source depend on the emissions levels of other precursors and / or sources?

Emissions may be non-uniform across sources, precursors or receptors, or any combination of the three. For example, greenhouse gas emissions are uniform across sources and receptors, but not across precursors (carbon dioxide, methane, etc.), each of which has a different global warming potential. When pollution problems are characterised by emissions that are uniform, linear and non-synergistic, the design of permit trading systems is relatively straightforward. In contrast, when emissions are non-uniform, non-linear and synergistic – as is the case with tropospheric ozone – the problems of designing an effective and efficient trading system are much more complex.

In a seminal paper Montgomery (1972) shows that, when emissions are linear and non-synergistic, a system of tradable pollution licences will induce the cost efficient allocation of emissions reductions across sources and precursors. Under *ambient permit trading* (APT), the authorities issue a quantity of *pollution permits* for each receptor point, equal to target pollution level at that point. Individual sources are then free to trade these permits, subject to the requirement that – after trading – they must hold sufficient permits for each receptor point to cover the pollution caused by their emissions of the various precursors. If the number of permits held by source i for receptor point k is denoted by q_{ki}, and the transfer coefficient for precursor m is given by α^m_{ki}, then this requirement can be formalized by the following K constraints:[4]

$$q_{ki} \geq \sum_m \alpha^m_{ki} e^m_i \quad (k=1,\ldots,K)$$

These constraints define the minimum portfolio of permits that the source must hold to support its mix of emissions e^m_i. Alternatively, they can be interpreted as

---

[4] The transfer coefficient α^m_{ki} gives the change in the ambient concentration of the pollutant at receptor point k for a unit change in emissions of precursor m at source i; e^m_i are the emissions of precursor m at source i

defining the allowable combination of emissions of the different precursors for a given portfolio of permits. While this "emissions rule" is relatively complex in the general case, it becomes much simpler as the uniformity of emissions increases. For example, if emissions are uniform across receptor points, then permits need only be issued for a single receptor point (i.e. the most restrictive one), and hence the emissions rule reduces to a single constraint. When emissions are uniform across sources, precursors and receptor points, then APT is completely equivalent to simple *emissions permit trading* (EPT). Under this system, the authorities determine the total amount of emissions that is consistent with the pollution objectives at each receptor point and issue a quantity of *emission permits* equal to this value, which can then be traded between sources on a one-for-one basis. The emissions rule for each source is now very simple – it must hold sufficient permits to cover its aggregate emissions across all precursors.[5]

The extension of APT to the cases of non-linear and synergistic emissions is analysed by Zylicz (1993). When emissions are non-synergistic he shows that APT will still induce the cost efficient outcome if the relationship between pollution levels and precursor emissions can be represented by a series of quadratic functions, although the emissions rule for each source must be amended accordingly. While this makes the rule slightly more complex than in the linear case, the extra computational burden is not that great. However when emissions are also synergistic, matters become much more complex. In this case, the emissions rule for each source is no longer independent from the emissions by other sources, and hence the optimal choice of permit holdings and emissions for each source now depends on the corresponding choices of the other sources. Consequently the outcome represents a Nash equilibrium of a two stage game. In the first stage, the optimal distribution of permits between sources is determined, together with the resultant (shadow) prices of the permits. In the second stage, the optimal allocation of emissions is determined for this distribution, with the price of permits acting as a co-ordinating mechanism for the equilibrium.

Unfortunately, this equilibrium will not – in general – coincide with the cost efficient outcome, and it is therefore necessary to supplement APT with a source specific tax, where the value of the tax payment for each source is dependent on the equilibrium permit price and the distribution of emissions between the sources. While this hybrid ambient permit-tax system will – in theory – induce a Nash equilibrium that coincides with the cost efficient solution, it is not at all clear what institutional mechanism could be used to ensure that sources obeyed their respective emissions rules when these are affected by other sources' emissions. This problem, together with the onerous computational requirements that it imposes, make it hard to imagine that such a system could be feasible in practice.

---

[5] See Tietenberg (1985) and Xepapadeus (1997) for detailed analyses emissions permit trading.

9.2.2
Ozone Control Problem

Fig. 9.1 illustrates the salient characteristics of the ozone control problem for the simple case of a single source that affects three receptor points (x, y and z). Ozone *isopleths* are shown for each receptor point. These contours give the combinations of precursor emissions for which the resultant ozone concentrations are exactly equal to the respective target values. For receptor x, precursor emissions are linear; ozone concentrations increasing with NO_X emissions. Hence, all combinations of emissions to the left of isopleth x produce ozone concentrations that are below the target value for this receptor point. Precursor emissions are also linear for receptor y, but for this point ozone concentrations decline as NO_X emissions increase. Consequently, only those combinations of emissions lying to the right of the isopleth are consistent with the target ozone concentration. Finally, at receptor point z precursor emissions are non-linear and / or synergistic, and the correlation between ambient ozone concentration and NO_x emissions switches from positive to negative as the level of emissions increases. In this case, the resultant ozone concentration is only less than the target value for combinations of precursor emissions lying below the u-shaped isopleth.

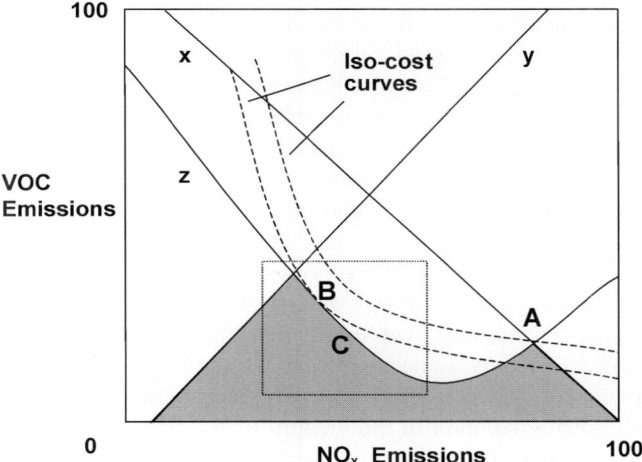

Fig. 9.1. Global ozone control problem

Together, the three constraints define a *feasible region* for the ozone control problem that is non-convex (shown as the shaded area).[6] The objective for the control problem is to find the combination of precursor emissions that will

---

[6] The feasible region gives the combinations of precursor emissions that produce ozone concentrations that are less than or equal to the target values at all three receptor points.

minimize total abatement costs while simultaneously satisfying the ozone target at all three receptor points (not necessarily with equality). This cost efficient solution occurs where the iso-cost contour is tangential to the boundary of the feasible region. If the feasible region is convex there will be a unique tangent-point.[7] However, when the feasible region is non-convex there may be many such points, each corresponding to a different value for total abatement costs. Only one (or possibly some) of these points will represent the true cost efficient solution (i.e. a global cost minimum); the others only being cost efficient in the vicinity of the tangent-point (i.e. a local cost minimum).

For the iso-cost curves illustrated in *Fig. 9.1* there are two solutions to the cost minimization problem – a global cost minimum at point A, and a local cost minimum at B.[8] Both of these points represent Nash equilibria under the hybrid permit-tax system, and there is no reason to suppose that the equilibrium corresponding to the global cost minimum would be chosen. Consequently, even if the hybrid system proposed by Zyliez (1993) could be made operational, it would not be guaranteed to yield the global cost minimum solution in all cases (although the outcome would always represent a local cost minimum).

From the preceding discussion it would appear that the prospects for using some form of permit trading to ensure the efficient implementation of ambient ozone objectives look bleak. However, if the expectations for a trading system are restricted to improving the cost efficiency of a particular initial allocation of emissions targets, rather than finding the global cost minimum solution from *any* initial allocation, then it may still be possible to design a workable approach. This objective is, of course, much less ambitious. However, it may often be all that is desired of an implementation mechanism; particularly in an international context, where sources are defined in terms of countries, and large-scale changes to the emissions targets specified in the initial agreement may not be politically acceptable.

If the initial allocation of national emissions targets is based on the results of cost optimization simulations (of the type described in the previous chapters) then it may be reasonable to assume that this is fairly close to a cost-minimizing solution, and hence that only some incremental fine-tuning is required. In this situation, all that is required of an implementation mechanism is that it be able to find the cost minimizing solution in the vicinity of the initial allocation (i.e. the nearest local cost minimum). An example of this, more limited role for emissions trading can be seen in the UN/ECE Second Sulphur Protocol. While the national emissions targets were based largely on cost minimizing scenarios produced by the RAINS model (Klaassen 1996), the Protocol allows for the possibility of joint implementation of emissions targets, presumably in recognition of the fact that these targets do not represent the true cost minimum solution.

---

[7] Assuming that the abatement cost functions are strictly convex.
[8] The value of total abatement costs associated with each iso-cost curve decreases as one moves towards the top right corner.

With this, more limited objective in mind, it is possible to construct a local, "linearized" optimization problem, centred on the initial allocation of emissions. Letting the relationship between ozone concentrations at receptor k and emissions of precursors from all sources (denoted by the vector e)[9] be defined by the function $w^k(e)$, then the Taylor approximation of order one around the initial emissions allocation $\mathbf{e}^0$ for this function is given by:

$$w^k(e) = w^k(\mathbf{e}^0) + \nabla w^k(\mathbf{e}^0) \cdot (e - \mathbf{e}^0) + R_1$$

Furthermore, if one assumes that in the neighbourhood of $\mathbf{e}^0$ (i.e. $N(\mathbf{e}^0)$) that the remainder term R_1 is close to zero, then:

$$w^k(e) \approx w^k(\mathbf{e}^0) + \nabla w^k(\mathbf{e}^0) \cdot (e - \mathbf{e}^0) \quad \text{for all } e \in N(\mathbf{e}^0)$$

Thus, for values of emissions close to the initial value $\mathbf{e}^0$, the relationship between ozone concentrations at each receptor point and precursor emissions can be approximated by a linear function, where the individual coefficients are given by the respective partial derivatives evaluated at the initial emissions level. This simplification is given some support by Simpson and Eliassen (1997), who note that, for long-term ozone statistics, there is a high degree of linearity in the system. Of course, the validity of the approximation will depend on the size of the neighbourhood $N(\mathbf{e}^0)$, but EMEP calculations suggest that it is likely to be reasonable for values of emissions within +/- 20% of the initial values.

Returning to the one source / three receptor example (see *Fig. 9.1*), the local optimization problem around the initial point C is shown in *Fig. 9.2*, where the shaded feasible region represents the "linearized" subset of the original feasible region about this point. In comparison with the original problem, it can be seen that the feasible region is now convex, and the solution B is a global cost minimum (albeit of this local problem). It should also be noted that one of the original receptor constraints is now redundant.

More formally, if the upper and lower bounds on the precursor emissions of source i are given by $\bar{e}_i^m$ and $\hat{e}_i^m$ respectively (where $m = N(O_x), V(OC)$), then the local optimization problem can be expressed as:

Minimize $\sum_{i \in I} C_i(e_i^N, e_i^V)$ subject to $\sum_{i \in I} (\alpha_{ki}^N e_i^N + \alpha_{ki}^V e_i^V) \leq w_k^*$ for all k

$$\hat{e}_i^N \leq e_i^N \leq \bar{e}_i^N \quad \text{for all } i$$

$$\hat{e}_i^V \leq e_i^V \leq \bar{e}_i^V \quad \text{for all } i$$

Since the constraint set is compact, a solution to this local problem is guaranteed to exist.[10] Furthermore, if the cost functions are all convex, then the

---

[9] To simplify the notation, superscripts distinguishing between the two precursors (NO$_X$ and VOCs), and source subscripts have been omitted.

[10] Weierstrass's Theorem: see Simon and Blume (1994), p. 823. It is assumed that the cost functions of each source are continuous.

problem is well behaved (i.e. the Kuhn-Tucker conditions are necessary and sufficient for a solution), and it is straightforward to show that the ambient permit system proposed by Montgomery (1972) can be amended to ensure that it yields the optimum outcome.

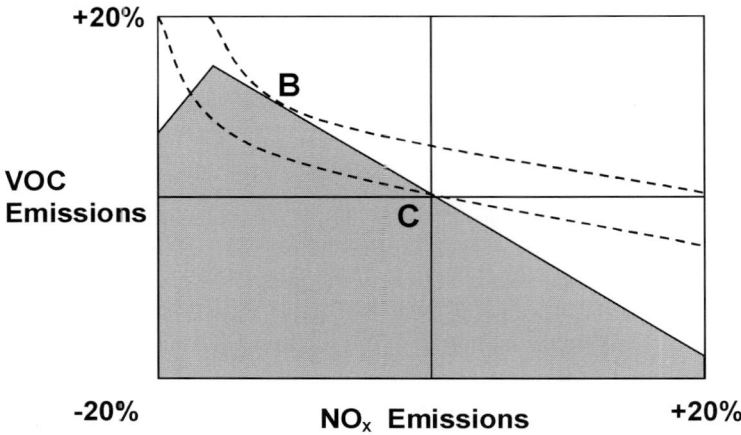

Fig. 9.2. Local "linearized" ozone control problem

However, doubts have been expressed about the cost efficiency of APT in practice. In particular, concerns have been raised in relation to the high transaction costs that are likely to arise under APT (Krupnick *et al.* 1983), and over the potential for strategic behaviour in the individual permit markets (Hahn 1986). Because sources must operate in multiple permit markets, the transaction costs associated with adjusting their emissions levels are likely to be relatively high, reducing the potential benefits of trading. Furthermore, since these costs can be expected to increase with the number of parties involved in the transaction (due to co-ordination problems, etc.), there is likely to be a reliance on bilateral trades rather than large scale multilateral deals. The overall effect will be to reduce the volume of permits traded (although not the number of transactions), and hence to prevent the attainment of the cost minimum outcome (Stavins 1995). A further problem can arise from the domination of receptor points by particular sources, due to geographical adjacencies or meteorological factors. This can lead to problems of market power in the permit markets (although this will depend on the initial allocation of permits), again preventing the attainment of an efficient outcome (Hahn 1984).

Both of these problems are directly related to the number of permit markets, and hence there may be a temptation to restrict the number of markets that are introduced to a small subset of potential receptor points. However, this would increase the likelihood of the environmental objectives being violated at those receptor points no longer covered by the permit system. Consequently, in practice there would appear to be an inevitable trade-off between the cost efficiency of APT and the its environmental effectiveness. In the context of international pollution problems, environmental effectiveness is likely to be of primary concern, and hence there will be a desire to maximize the number of receptor points. If this is the case then there are strong reasons to believe that APT will fail to deliver the cost efficient outcome.

It would appear therefore that for pollution problems that are non-uniform, non-linear, or synergistic, "pure" permit trading systems will not be able to reconcile the desire for cost efficiency with the need for environmental effectiveness. Consequently, the next section will consider an alternative, hybrid approach; one that combines limited permit trading with administrative procedures, in order to improve cost efficiency while ensuring a high level of environmental effectiveness at all receptor points, at all times.

9.2.3
Iterative Zonal Permit Trading

Iterative zonal permit trading (IZPT) represents a development of the well established concept of *zonal permit trading* (ZPT)[11], under which a control region is divided into a number of separate zones, and emissions budgets (or targets) are set for each zone.[12] Within each of the zones, individual sources are then allowed to trade emissions on a one-for-one basis, but trading between zones is prohibited. The key extension of IZPT is the recognition that the price information provided by the individual permit markets can itself be used to revise the allocation of emissions budgets in an iterative administrative process that will improve overall cost efficiency of the abatement programme. While the approach is applicable to any situation where pollution levels are sensitive to the location of emissions, it has a number of features that make it particularly relevant for international transboundary environmental problems such as tropospheric ozone:

- it can incorporate a large number of receptor points, and it ensures that ambient air quality targets at each receptor point are met at all times;
- it converges towards the least cost solution, and reduces aggregate abatement costs at each revision (except possibly the final revision);
- it reveals information about the changes in abatement costs for each zone;
- it facilitates burden-sharing on the basis of the net abatement costs arising under the final allocation of emissions budgets to zones;
- it is relatively simple to implement compared to the alternative international permit trading systems;

---

[11] See Tietenberg (1985) for a full discussion of zonal permit trading.
[12] The definition of zones is very flexible, and can allow a number of different interpretations. In the particular context considered in this chapter, zones are defined in terms of countries and precursors (e.g. zone 1 = United Kingdom / NO_x , zone 2 = United Kindom / VOC, etc.).

- it represents only an incremental change to the existing institutional framework for international agreements (e.g. the UN/ECE Second Sulphur Protocol).

The original concept of ZPT is based on the premiss that within each zone there is (at least approximately) uniform mixing of emissions, i.e. the location of the emissions source within a particular zone does not matter. If this is the case, then the transfer coefficients for each emissions source in a particular zone ($g = 1,\ldots G$) are identical (i.e. for all $i \in I_g$, $\alpha_{ki} = \beta_{kg}$)[13], and hence – in the optimal solution – the marginal abatement costs of all sources within the zone are equalised. This observation allows the original optimization problem to be decomposed into two, linked optimization problems:

Minimize $\sum_{g \in G} C_g^{z*}(e_g^z)$ subject to $\sum_{g \in G} \beta_{kg}\, e_g^z \leq w_k^*$ for all $k \in K$

$$\hat{e}_i^s \leq e_i^s \leq \overline{e}_i^s$$

Minimize $\sum_{i \in I_g} C_i^s(e_i^s)$ subject to $\sum_{i \in I_g} e_i^s \leq e_g^z$ for all $i \in I_g$

Where e_g^z represents the aggregate emissions for zone g, and $C^{z*}_g(e_g^z)$ is the optimized aggregate cost function for zone g, which is given by the value function for the second optimization problem.[14] The first problem relates to the determination of optimal emissions budgets for each zone, given that the allocation of this budget among individual sources within the zone is cost efficient. The second problem relates to the efficient allocation of any given zonal emissions budget between its constituent sources. Since for this problem emissions are uniform across sources, it is possible to use EPT as the implementation mechanism. Because this is much simpler than APT, it is less prone to suffer from high transaction costs. Also, within each zone there are likely to be a large number of individual sources, minimizing the potential for strategic behaviour. Consequently, EPT can be expected to produce an outcome that is reasonably close to the optimal allocation of emissions within the zone.

However, Tietenberg (1985) raises two major concerns about ZPT which are likely to undermine its cost efficiency in practice. The first relates to the assumption of homogeneous transfer coefficients within a particular zone, and hence to the efficiency of trading emissions rather than ambient concentrations. To the extent that transfer coefficients are not homogeneous, one-for-one trading of emissions between sources may lead to the creation of pollution "hotspots". If this

---

[13] The notation I_g is used both as a label for the subset of sources belonging to zone g, and also to represent the total number of sources in the zone. Hence, $I = I_1 + I_2 + \ldots + I_G$ under either interpretation. Similarly, G is used both to represent the set of zones, and also the number of zones. The relevant interpretation should be clear from the context.

[14] The superscripts s and z denote whether the emissions and cost functions relate to individual sources, or to zones.

problem is to be avoided, then the emissions budget for the zone will have to be reduced from its efficient level, increasing the overall compliance cost for that zone. When zones represent large geographical areas – such as countries – the validity of common transfer coefficients must be open to question. However, practical considerations mean that the assumption is common in the modelling of transboundary pollution[15], and it is implicit in the setting of national emissions targets.

The second concern is much more serious, and it relates to allocation of emissions budgets between zones. As can be seen from the two optimization problems above, the efficient allocation of emissions budgets to zones requires complete knowledge of the abatement cost functions of each individual source (in order to calculate the zonal cost functions). Clearly, this is not a realistic expectation. Indeed if it were, it begs the question as to why emissions targets could not be calculated directly for individual sources. While emissions budgets may be calculated on the basis of estimated cost functions for the different zones, it is highly unlikely that the outcome will represent the true cost efficient allocation. Because ZPT does not allow trading between zones, the cost inefficiencies of the initial allocation are "locked in" to the system. Tietenberg (1985) views this inability to correct the inevitable mis-allocation of emissions budgets as a serious handicap for ZPT compared to alternative trading approaches.

IZPT addresses the second of these concerns. In particular it provides a mechanism for revising the initial allocation of emissions budgets – based on the prices of emissions permits in each zone – so as to reduce the initial cost of the control programme. The basic foundation for the approach is provided by the observation that, for the prevailing allocation of emissions budgets $e^z = (e_1^z, \ldots e_G^z)$, not only is the equilibrium market price of permits in each zone equal to the marginal abatement cost of each individual source, but also it is equal to the derivative of the value function $C^{z^*}_g(e_z^g)$, i.e. the marginal abatement cost for the zone:

$$p_g = p_g\left(e_g^z\right) = -C_g^{z*\prime}\left(e_g^z\right) = -C_i^{s\prime}\left(e_i^s\right) \qquad \text{for all } i \in I_g$$

Consequently, the gradient vector $\nabla \mathbf{C}$ for the iso-cost surface

$$C_1^{z*}(e_1^z) + \ldots + C_G^{z*}(e_G^z) = \overline{C}$$

is equal to the (negative of the) permit price vector, i.e. $-\mathbf{p} = (-p_1, \ldots, -p_G)$. Thus, while the iso-cost surface itself is not directly observable, its gradient vector at the current allocation of emissions budgets can be derived indirectly from the permit prices in each zone. By comparing this vector with the boundary of the

---

[15] For example, the INFOS modelling has been based on country-to-grid relationships between precursor emissions and ambient ozone concentrations.

feasible region it is possible to determine, not only whether the current allocation is optimal, but also in which direction to move in order to improve cost efficiency, if it is not.

The basic concept of the revision process is illustrated in *Fig. 9.3*, for the simple case of two zones. The initial emissions budgets for the two zones are denoted by the point D, while the point B represents the cost efficient allocation. The gradient vector ∇C for the (unknown) iso-cost curve at point D is given by the (negative of) the equilibrium permit prices for each zone $\mathbf{p} = (-p_1, -p_2)$. For any direction vector $\mathbf{d}$ along the boundary of the feasible region, if the angle between the vectors $\mathbf{d}$ and $\mathbf{p}$ is greater than 90 degrees[16], then a move in that direction will decrease aggregate costs. Conversely, if it is less than 90 degrees then moving in that direction will increase costs. It follows directly that if the angle between the two is less than (or equal to) 90 degrees for all possible direction vectors, then the current allocation is cost efficient. In this example there are two possible direction vectors along the boundary of the feasible region ($\mathbf{d}_1$ and $\mathbf{d}_2$), and it is clear that only a move along $\mathbf{d}_1$ will reduce aggregate costs.

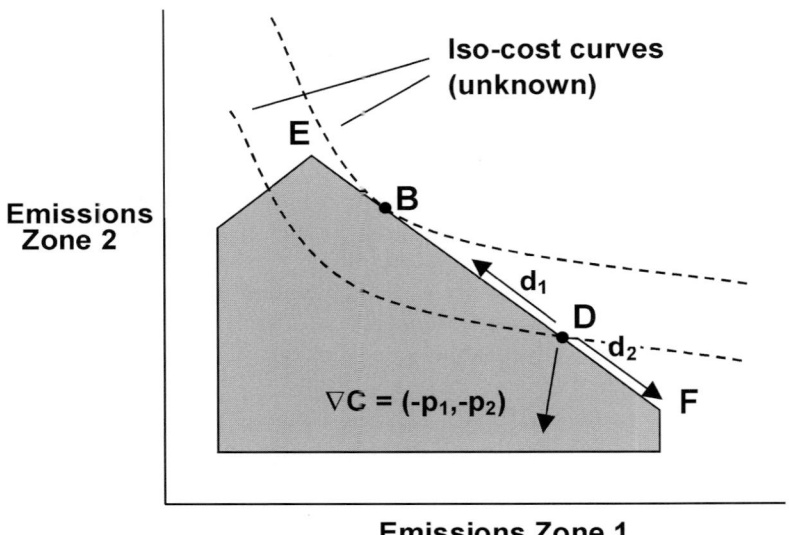

Fig. 9.3. Concept of iterative zonal permit trading

Unfortunately, while this simple decision rule tells us in which direction to move in order to reduce costs, it does not tell us how far to move. One possible *revision algorithm* would be to move as far as possible in the identified direction, while remaining on the feasible surface (i.e. to point E in *Fig. 9.3*). However, as

---

[16] More formally, if the inner product of the two vectors is negative.

can be seen, this is likely to lead to the cost minimum solution being "overshot", and to the possibility that the resultant change in emissions budgets could actually increase aggregate costs. Furthermore, if the algorithm were then to be applied again it would result in the emissions budgets being revised to point F, and subsequently to oscillation between these two extreme points on the feasible surface.

In order to minimize the overshoot problem, arbitrary *revision bounds* must be set for each zone – denoted by the vector $\overline{\mathbf{d}}$ – which define the maximum distance that can be moved in any particular direction (i.e. the maximum allowable change to each zone's emissions budget). The revision algorithm can then be applied iteratively, moving the emissions budgets towards the cost minimum solution B in a step-by-step process. While this amendment does not eliminate the possibility of the solution being overshot, the setting of relatively "tight" revision bounds will greatly reduce the scale of the problem. Of course, oscillation around the cost minimum solution can still occur – albeit over the much smaller range defined by the revision bounds.[17]

To deal with the problem of oscillation, the revision algorithm must be supplemented by a rule that defines when the iterations should stop. One simple rule would be to halt the process if a move in the same direction as that determined at the previous iteration would increase aggregate costs. For example in *Fig. 9.3*, if the revision bounds had been sufficiently large to allow a move from D to some point in the interval BE, then at this point the angle between $\mathbf{d_1}$ and $\nabla \mathbf{C}$ would now be less than 90 degrees. Hence a further move in this direction would increase aggregate costs. Of course this rule requires that the solution must be overshot before the process ceases, and consequently the proximity of the final budget allocation to the optimal solution will depend on the size of the revision bounds.

These basic concepts can be formalised in an iterative algorithm that can be used to generate revisions to emissions budgets for the general case of G zones. The algorithm represents an adaptation of the "feasible direction algorithm" which is sometimes used in the numerical solution of non-linear programming problems.[18] The vector of revised emissions budgets at a particular iteration r (i.e. $\mathbf{e}^r$) is given by the mapping $A^r(\mathbf{e}^{r-1}) = \mathbf{e}^{r-1} + \mathbf{d}^r$, where the vector $\mathbf{d}^r$ is the solution to the following linear programming problem:

Minimize $\quad \nabla \mathbf{C}(\mathbf{e}^{r-1}) \cdot \mathbf{d} \quad$ subject to $\quad \mathbf{B} \cdot \mathbf{d} \leq \mathbf{w}^* - \mathbf{B} \cdot \mathbf{e}^{r-1} \quad |\tilde{d}| \leq \overline{\mathbf{d}}$

where $\nabla \mathbf{C}(\mathbf{e}^{r-1}) = -\mathbf{p}^{r-1}$. The first constraint ensures that the revised emissions budgets remain feasible, taking account of any slack applying at the current emissions budgets, while the second relates to the problem of overshooting discussed above.

---

[17] If for some iteration, the revised emissions budgets happen to coincide with the cost minimum solution then the process will come to a stop, and the problems of overshoot and oscillation will not arise.

[18] See Luenberger (1965).

The theoretical properties of this algorithm have been established by Salmons (1999). For a given vector of revision bounds $\bar{d}$, and an initial vector of emissions e^0, the algorithm will generate a sequence of emissions budgets that converge to a solution point (e^{**}) in a defined solution set $\Gamma_{\bar{d}}$, where the "size" of this solution set depends on the values that have been set for the revision bounds. Furthermore, as the revision bounds for each zone tend to zero (i.e. $\bar{d} \to 0$), the solution set $\Gamma_{\bar{d}}$ shrinks to the cost minimizing solution e^*, and hence the final cost under the algorithm tends to the true cost minimum (i.e. $C(e^{**}) \to C(e^*)$).

Thus, by choosing smaller values for the revision bounds, the algorithm will yield a more precise approximation of the true solution. However, in most cases, this will be at the expense of requiring an increased number of iterations to reach the solution. While this trade-off is of no consequence for theoretical simulations where adjustments occur instantaneously, it represents a significant design issue for the approach in practice. Given the length of time that it is likely to take for the revisions produced by the algorithm to be ratified, and for the permit markets in each zone to reach their new equilibria, there will need to be a reasonably long interval between successive revisions (e.g. 3–5 years). Thus, although the setting of tighter revision bounds would lead to greater cost efficiency in the long run, it is likely to lead to higher costs in the short-medium term (see *Fig. 9.4*). Depending on the relative cost profiles over time, and the choice of discount rate, it is possible that the net present value of costs could be lower for the larger value of the revision bounds.

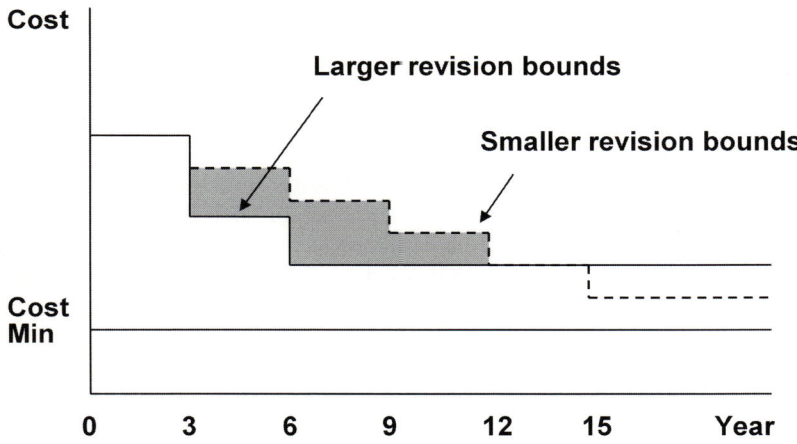

Fig. 9.4. Cost profiles over time for different revision bounds

It might be inferred from *Fig. 9.4* that, for discount rates above a certain level, it would always be preferable to choose a relatively large value for the revision bounds in order to maximize the cost savings achieved at the first iteration. However, this is not necessarily the case. If the initial allocation of emissions budgets is close to the true cost minimum solution, then the setting of large revision bounds will lead to the solution being overshot at the first iteration, with the possibility that costs could actually increase (see *Fig. 9.3*). Of course, since the cost profiles are not known *a priori*, the choice of revision bounds must inevitably be subjective. In general however, the closer the initial allocation is believed to be to the true cost minimum solution (i.e. the better the initial approximation), the tighter the revision bounds should be set.

As the algorithm moves the budget allocation towards the optimal solution it is possible that the emissions budget for a particular zone may be increased at one iteration, only to be decreased at the next (or subsequent) iteration. Again, while this "zigzagging" may be of no theoretical consequence, it is likely to have serious implications for the feasibility of IZPT in practice. While the time interval between the revisions may be reasonably long, the possibility that changes to emissions budgets may be reversed at subsequent revisions introduces an additional element of uncertainty into the planning of long term investment decisions. One simple way of avoiding the problem would be to prohibit revisions to emissions budgets from changing sign from one iteration to the next. Thus, if the emissions budget for a particular zone is increased (decreased) at the first iteration, it cannot be decreased (increased) at subsequent iterations. However, it should be recognised that while this additional constraint may improve the political feasibility of the revision algorithm, it may also lead to a situation where the solution set $\Gamma_{\bar{d}}$ no longer contains the optimal vector $\mathbf{e}^*$. Consequently, it may no longer be the case that the final cost under the algorithm tends to the minimum cost as the revision bounds tend to zero.

In order to ensure that the overall allocation remains within the feasible region at each iteration, any increases in emissions budgets for some zones must be offset by reductions in the emissions budgets of others. Hence, the revision process will necessarily create winners and losers; some zones enjoying reductions in abatement costs while others face cost increases. While this may not be too problematic in a national setting, where the authorities can impose the revisions unilaterally, it is extremely unlikely that the revisions could be implemented in an international context – where zones are defined in terms of countries – unless they are accompanied by explicit financial transfers to compensate the losers.

Unlike permit trading, where the revenue from the permit transactions provides automatic financial compensation for increased abatement effort – any financial transfers to accompany the revisions to the emissions budgets must be calculated separately. This has the advantage of allowing the introduction of "benefit-sharing", under which the aggregate benefit (i.e. reduction in total costs) at each iteration is shared between the zones according to some equity principle.[19]

---

[19] The relationship between benefit-sharing and burden-sharing – which applies to the cost of the initial emissions budget allocation – is discussed in *Sect. 9.3.1*.

However, it requires information on the scale of the gains and losses of the individual zones. Clearly, if the calculations were to rely on the self-reporting by the individual zones (i.e. countries), there would be a strategic incentive for them to overstate true increases in costs and to understate true decreases in costs, in order to influence the size of the financial transfer that they must pay, or that they receive.

Fortunately, one of the useful features of IZPT is that it provides direct, objective information on the scale of the cost saving at each iteration – both in aggregate, and for the individual zones. Given the underlying assumption that the cost function of each zone is convex, the aggregate cost function is also convex, and it follows directly from the definition of a convex function that

$$\nabla C(e^r) \cdot (e^r - e^{r-1}) \geq C(e^r) - C(e^{r-1}) \geq \nabla C(e^{r-1}) \cdot (e^r - e^{r-1})$$

However, since the cost gradient vector ∇C is equal to the negative of the permit price vector **p**, this double inequality can be restated as

$$\mathbf{p}^{r-1} \cdot \mathbf{d}^r \geq C(e^{r-1}) - C(e^r) \geq \mathbf{p}^r \cdot \mathbf{d}^r$$

The middle term is the actual cost saving resulting from the changes made to the emissions budgets at iteration r. The left hand term is the value of the change in emissions budgets based on the permit prices prevailing before the change (i.e. the *ex ante* valuation), while the right hand term represents the value based on the prices prevailing after the change has been made (i.e. the *ex post* valuation). Respectively, these two estimates provide upper and lower bounds on the actual cost saving that is achieved. Thus while the actual cost saving arising from a change to emissions budgets is not observable, it is possible to calculate an interval in which it will lie based on the permit prices in each market. By the same argument, it is also possible to construct equivalent intervals for each zone, enabling estimation of the individual cost reductions, or cost increases.

From a practical standpoint, two of the major attractions of IZPT are its relative simplicity, and the fact that it does not represent a major departure from the current approach to international environmental agreements. For each precursor, national emissions targets are set and these are only changed by an administrative procedure, allowing political scrutiny of the process at each revision. Given information on the permit prices for each zone, and the transfer coefficients between the zones and the receptor points, it is straightforward to calculate the recommended revisions to the emissions budgets. The revision algorithm can be set up on a simple spreadsheet, and the calculations take a matter of seconds to complete. In the context of a European Union abatement programme, responsibility for collecting the information, and for performing the calculations might rest with DG XI (Environment, Nuclear Safety and Civil Protection), or with the European Environment Agency. While it is possible that the initial agreement might allow the recommended revisions to be implemented automatically, it is more likely that they would have to be ratified by the European Parliament, and the Council of Ministers, prior to implementation. This political "check and balance" would allow a degree of negotiation over the revised emissions budgets.[20] It would also allow for the agreement of the rules for calculating the benefit-sharing transfer payments that would accompany the revisions.

In concluding the discussion of IZPT, it is interesting to compare it with three alternative approaches:

- zonal emissions budgets with an emissions tax in each zone;
- zonal emissions budgets with individual emissions targets set for each source;
- ambient permit trading (APT) between zones.

In theory it would be possible to use zone-specific *emissions taxes*, instead of EPT, to achieve each zone's emissions budget at least cost (Baumol & Oates 1971). In this case, the tax rates for each zone would provide the necessary information for the revision algorithm. However, in the absence of perfect information about the abatement costs of each source, it is highly unlikely that the initial values of the tax rates would be set at the optimal level. Therefore the rates would need to be adjusted in an iterative process until the budgets were achieved. This has two major disadvantages compared to EPT. Firstly, if any of the tax rates are set below their optimal levels then the respective emissions budgets will be exceeded during the adjustment process, and hence ambient air quality targets will be violated at some of the receptor points. Secondly, the tax adjustment process may take much longer to complete than the permit market would take to reach equilibrium under EPT; requiring a number of iterations, with any tax increases facing strong political opposition from the effected sources. Consequently, the interval between revisions of the zonal emissions budgets – and the resultant cost saving – is likely to significantly longer when emissions taxes are used to implement the targets.

Alternatively, the authorities responsible for each zone may wish to set individual *emissions limits* for their constituent sources.[21] While these targets may be based on a view of the relative abatement costs, it is unlikely that they would represent the true cost efficient allocation of abatement effort within the zone. However, the major problem with this approach is that it does not provide any information about the shadow values of the emissions constraints for the zones (i.e. the marginal costs of abatement for the zones). Without this information, it is not possible to determine how the zonal emissions budgets should be revised in order to reduce total abatement costs.

In theory, both IZPT and APT have the potential to induce the cost minimizing allocation of abatement effort between zones. While they may fail to reach this objective in practice – due to transaction costs and strategic behaviour in the case of APT, and due to the choice of finite revision bounds in the case of IZPT – each

---

[20] A precedent for this is provided by the negotiations that took place over the emissions targets recommended by the RAINS model during the formulation of the UN/ECE Second Sulphur Protocol (Klaassen 1996).

[21] In the international context considered here, the responsible authorities would be national governments (or their agencies).

will improve the cost efficiency of the initial allocation. However, there are a number of major differences between the two approaches.

As was noted in *Sect. 9.2.2*, the number of receptor points (i.e. permit markets) that can be accommodated under APT is likely to be fairly limited in practice. In contrast, because revisions are implemented by an administrative procedure, there is no limit to the number of receptor points that can be included in IZPT. For example, it would be perfectly possible to include feasibility constraints in the revision algorithm for several hundred EMEP grid squares if desired. While this might reduce the scale of the potential cost reductions, it would not affect the ability of the algorithm to produce the optimal budget revisions. Consequently, the environmental effectiveness of IZPT is likely to be much greater than that of APT, with the possibility of pollution hotspots being greatly reduced.

While the automatic financial transfers provided by the permit transactions under APT ensure that all parties are better off after the exchanges, IZPT requires the separate calculation of transfer payments to compensate the losers at each revision. However, this has certain advantages in relation to the calculation of burden-sharing transfers. As will be explained in *Sect. 9.3.1*, the calculation of burden-sharing transfers on the basis of the net abatement costs arising under the final allocation of emissions budgets (including any financial transfers from permit transactions) can be decomposed into two parts – burden-sharing transfers based on the gross abatement costs at the initial allocation, plus benefit-sharing transfers based on the reduction in costs in moving from the initial to the final allocation. The benefit-sharing transfer is calculated according to the following formula:

$$\text{Benefit Sharing Transfer}_g = \rho_g \times \text{Aggregate Benefit} - \text{Individual Benefit (Cost)}_g - \text{Financial Transfer}_g$$

where the share ρ_g reflects the underlying equity principle (see *Sect. 9.3.1*). Unless the financial transfer arising from the permit transactions under APT exactly offsets the difference between the imputed benefit and the actual benefit (cost), then some additional benefit-sharing transfer will be required. However while the value of the permit transactions is known, APT does not provide any information on the benefits arising from the changes. In contrast, as we have seen above, IZPT provides all of the necessary information to calculate the benefit-sharing transfer payments for any value of ρ_g.

The final difference relates to the length of time that will be required to reach the final allocation of emissions budgets. In theory one would expect APT to reach its final equilibrium more quickly, since transactions take place in a continuous process, while emissions budgets are only revised periodically under IZPT. However, in practice the comparison between the two systems is not so clear. On the realistic assumption that APT would operate via a series of small scale, sequential transactions – with a preponderance of bilateral transactions – then unless there is some co-ordinating institutional mechanism, it is not possible to predict the time profile of aggregate costs under APT. This makes it difficult to make any definitive *a priori* judgement about the relative speed of convergence of the two systems. However, the greater the number of potential trading partners, the longer APT is likely to take to reach its equilibrium, and hence the less likely it is to converge more quickly than IZPT.

9.2.4
Illustrative Example

In this section, a simple illustrative example is used to show how IZPT can be used to "fine tune" an international ozone abatement programme, so as to improve its cost efficiency. In this example there are three countries (imaginatively called A, B and C), each emitting both NO_x and VOCs (i.e. there are six zones). For each zone, the relationship between abatement costs (C) and emissions (e) takes the same generic form:

$$C = C_0 \cdot \left(1 - \frac{e}{\bar{e}}\right)^\gamma$$

where $\bar{e}$ is the pre-abatement level of emissions, C_0 is the (finite) cost of reducing emissions to zero, and γ determines the curvature of the cost curve; parameter values being chosen to produce the variation in marginal abatement cost curves shown in *Fig. 9.5*.[22]

There is one "average" receptor point in each country, and the target ozone concentrations at these points are 100, 150 and 50 respectively. The "local" source-receptor matrix in the neighbourhood of the modelled solution is given in *Table 9.2*. For all three countries, increases in emissions of VOC's have a positive impact on ozone concentrations in all countries (apart from country A, which has no impact on country C). Increases in NO_x emissions by country B lead to an increase in ozone concentrations in all three countries. However for countries A and C, NO_x emissions are negatively correlated with ozone concentrations in country C.

---

[22] Neither the cost curves, nor the source-receptor matrices used in this example are intended to reflect those used for the optimization modelling described in the preceding chapters. They have been chosen so as to accentuate particular features of IZPT in order to facilitate understanding. Consequently, the resultant cost figures are purely hypothetical, and bear no relationship to those provided in *Chap. 7*.

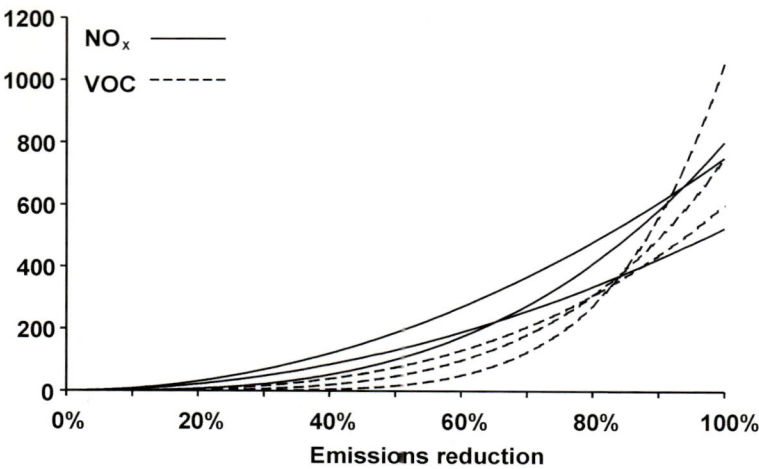

Fig. 9.5. Marginal abatement cost curves

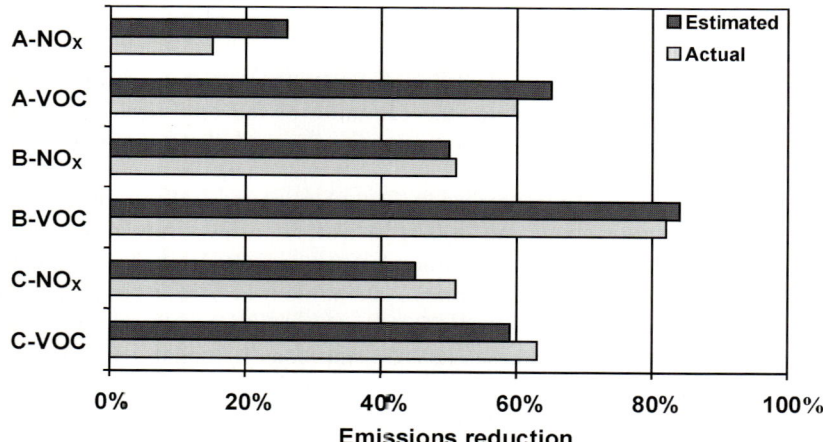

Fig. 9.6. Comparison of estimated and actual cost minimum solutions

Table 9.2. Source-receptor matrix

	NO_x			VOC		
Δ Ozone	A	B	C	A	B	C
Country A	0.20	0.05	0.00	0.60	0.20	0.10
Country B	0.00	0.30	0.30	0.20	0.80	0.15
Country C	-0.10	0.15	-0.20	0.00	0.10	0.60

The maximum revision bound for each zone was set at 5 % of its initial emissions budget, and the additional constraint on the signs of the revisions was included to prevent "zigzagging". *Table 9.3* shows the revised emissions budgets after each iteration. The algorithm converged after three iterations, with no further changes being feasible at the fourth iteration. For five of the zones, the process moves the emissions budget closer to the true cost minimum solution. However, in the case of NO_x emissions by country B, the algorithm moves the budget further away from its cost efficient value. This is due to the restriction on the sign of the changes at each iteration, and it illustrates clearly how the inclusion of this additional constraint can prevent the attainment of the cost minimum solution.

Table 9.3. Evolution of emission targets

	Country A		Country B		Country C	
	NO_x	VOC	NO_x	VOC	NO_x	VOC
Est. cost minimum	111.6	70.9	150.3	43.0	136.4	102.7
Iteration 1	117.2	74.4	151.6	45.2	129.6	97.5
Iteration 2	122.8	77.9	151.6	47.3	122.8	95.0
Iteration 3	128.4	79.4	151.6	49.5	118.4	89.9
Iteration 4	128.4	79.4	151.6	49.5	118.4	89.9
Act. cost minimum	126.8	80.5	147.4	48.9	122.7	91.7

The reduction in aggregate abatement costs over the three iterations was 9,100 million EURO, representing 98% of the potential cost improvement, despite the inclusion of the constraint on the sign of the revisions. The first iteration alone achieved almost half of the potential reduction in costs. Of course, the actual cost savings would not be observable in practice. However, as can be seen in *Table 9.4*, taking the simple average of the upper and lower bounds for the estimated cost savings – which as noted in *Sect. 9.2.3* can be calculated from the permit prices – provides a close approximation of the actual cost saving at each iteration.

Table 9.4. Cost savings

	Estimated cost saving				
	Upper	Lower	Average	Actual saving	Cum. % saving
Iteration 1	5 272	3 834	4 553	4 552	49 %
Iteration 2	3 839	2 605	3 222	3 219	83 %
Iteration 3	1 848	889	1 369	1 367	98 %
Iteration 4	0	0	0	0	98 %

However, while aggregate abatement costs are reduced at each iteration, the impacts on individual countries vary enormously. *Table 9.5* shows the winners and losers as a result of the revisions to the emissions budgets at the first iteration. Countries A and B both enjoy reductions in their total abatement costs – reflecting the increases in their respective budgets, although for country A this is accompanied by a slight increase in ozone concentrations. In contrast, the revisions impose a substantial cost increase on country C (up 27 % on the initial level). While this additional cost burden may be offset to some extent by the improvement in air quality, it is unlikely that country C would agree to the revisions being implemented unless they received some compensating transfer payment from the other two countries. Of course, this is just a hypothetical example, but it highlights the need for the revisions to emissions budgets to be accompanied by some explicit benefit sharing scheme.

Table 9.5. Distributional impact (iteration 1)

	Estimated cost saving			Actual saving	% Change in Ozone
	Upper	Lower	Average		
Country A	4 688	3 967	4 328	4 320	+1.3 %
Country B	3 533	3 415	3 474	3 474	0 %
Country C	-2 950	-3 547	-3 249	-3 242	-1.5 %
Total	5 272	3 834	4 553	4 552	

9.3
Equity and Burden-Sharing

In this second part of the chapter, the related issues of equity and burden-sharing are considered in detail. The discussion is divided into four sections. In the first, a number of "focal" equity principles are identified, and a pragmatic approach for balancing the alternative principles is described. This is followed by the definition of an institutional mechanism that could be used to implement a burden-sharing scheme in the particular case of tropospheric ozone pollution. The third section considers how the general principles can put into operational form in order to calculate burden-sharing transfers in relation to a particular INFOS reduction scenario. Finally the results of the analysis for the different approaches are compared and discussed.

9.3.1
Underlying Principles

The issue of equity is not new; it has been the focus of philosophical and political thought since ancient times. However, the role and importance of the equity in relation to environmental policy – and the associated issue of burden-sharing – has been the subject of increasing debate in recent years.[23] The main focus of this debate has been on how the cost burden of reducing greenhouse gas emissions should be distributed between nations[24], but a number of authors have also considered the issue of burden-sharing in relation to European "transboundary" pollutants such as SO_2 and NO_x (Sliggers and Klaassen 1994; Atkinson 1998).

Unfortunately, unlike the issue of efficiency, there is no universal consensus regarding the appropriate definition of equity, at either the international or interpersonal level. This has led to the adoption of a range of different philosophical and policy approaches, which have resulted in a degree of confusion in the debate. In particular, Rose (1992) notes that there is often a confusion between the underlying equity principle and the "reference base" which is used to evaluate a policy decision against that principle. For example, while population provides the relevant reference base for determining an initial allocation of emissions permits under the "egalitarian" principle that all people have an equal right to pollute, it is also consistent with a number of other principles.

A further potential source of confusion arises from the fact that the fairness of a policy initiative can be assessed in a number of different ways. Banuri et al (1996) make a distinction between "process" (i.e. procedural issues) and "outcomes" (i.e. consequentialist issues). The former is concerned with whether the process by which a decision is reached is fair (i.e. all parties are allowed to participate, and are treated equally), while the latter is concerned directly with the distribution of costs and benefits. Rose et al (1998) make a further distinction between "process", "outcomes" and "entitlements". Burtraw and Toman (1992) argue that any agreement which is entered into voluntarily in a process that is open and fair should be considered as being equitable. However, despite this view the majority of the analyses that have addressed the issue have considered equity from the standpoint of either the outcomes, or entitlements of the parties.

Closely associated with this is the issue of the appropriate "object" of the equity analysis (Rose 1992), i.e. what exactly is being distributed.[25] There are three possible bases for the calculations: the *ex ante* gross abatement costs prior to any joint implementation activities; the *ex post* net abatement costs after joint implementation (including any associated financial transfers); the *ex post* net welfare changes, taking account of any environmental benefits that are attributable to the policy initiative.

Following a welfare economics paradigm, ideally one would wish to base the analysis on the net welfare changes of each country. However, in practice most authors have focused on the distribution of costs. There are a number of reasons for this. Firstly, there is often a great deal of uncertainty – and controversy – surrounding the valuation of environmental improvements. Secondly, it may not

---

[23] Banuri et al (1996) provide an excellent overview of the subject in their contribution to the Second Assessment Report of the Intergovernmental Panel on Climate Change (IPCC).
[24] See Rose et al (1998) for a review.
[25] Of course, this issue only arises when equity is being assessed from an "outcomes" or an "entitlements" perspective.

be possible to place a monetary value on all of the improvements that occur – or even to quantify them in physical terms. A related problem arises in the case of tropospheric ozone, where reductions in precursor emissions will generate significant "primary" benefits in addition to the "secondary" benefits arising from the reduction in ozone concentrations. Indeed it only really makes sense to aggregate costs and benefits in an integrated *multi-pollutant, multi-effect* framework, where all of the primary and secondary impacts can be taken into account. Consequently, the ensuing discussion (and the analysis in *Sect. 9.3.3*) follows the traditional path of concentrating on distribution of abatement costs only.

If equity is being assessed from an "outcomes" perspective, then it is clearly desirable to calculate any burden-sharing transfer payments on the basis of the *ex post* net abatement costs. However, this is complicated by the need to estimate the impact of any joint implementation activities that may be undertaken by the parties (e.g. permit trading) – both in terms of the resultant changes in abatement costs, and in terms of the associated financial transfers. Fortunately, it is straightforward to show that the *ex post* burden-sharing transfers can be decomposed into two component parts:

Ex post burden-sharing transfer (net)	=	*Ex ante* burden-sharing transfer (gross)	+	*Ex post* benefit-sharing transfer (net)

This allows the burden-sharing process to be split into two stages. In the first stage, burden-sharing transfers can be calculated by applying the desired equity principle to the gross cost burden prior to any joint implementation. In the second stage, benefit-sharing transfers are calculated by applying the same equity principle to the aggregate reduction in abatement costs arising as a result of the joint implementation activities, taking explicit account of any financial transfers that may have occurred. The issues associated with the calculation of benefit-sharing transfers have been discussed in *Sect. 9.2.3* in relation to IZPT. The remainder of the chapter therefore focuses on the first stage – the calculation of burden-sharing transfers on the basis of the gross costs arising under the initial allocation of abatement effort.

According to the dictionary definition, equity means "something that is fair and just" (Flexner 1987). As such, an assessment of the equity of a particular policy initiative represents a normative evaluation, or value judgement, regarding the desirability of the distribution of burdens and benefits. This leaves the issue open to a variety of interpretations, and inevitably this has led to a number of different principles being proposed, with varying degrees of relevance to different environmental problems. For example, Rose (1992) identifies ten alternative equity criteria in relation to global warming policy. However, there are a number of recurrent themes that are relevant across a broad range of situations. Burtraw and Toman (1992) define these as focal points[26], and Atkinson *et al* (1999) suggest three of these as being particularly pertinent:

- burdens should be allocated on the basis of who is responsible for the problem (i.e. *polluter pays*);
- burdens should be allocated on the basis of who benefits from the removal (or reduction) of the problem (i.e. *beneficiary pays*);
- burdens should be allocated on the basis of financial means (i.e. *ability to pay*).

The first two approaches can be viewed as different interpretations of the sovereignty principle put forward by Rose *et al* (1998), where the interpretation is dependent on the distribution of property rights between the polluter and the victim (Barzel 1997). According to the sovereignty principle, "all nations have an equal right to pollute and to be protected from pollution".[27] If the established property rights regime puts more emphasis on the countries' rights to pollute, then it is the second approach that is likely to be dominant. In contrast, if it places greater weight on the countries' rights to be protected from pollution, then the first approach is more relevant.

Responsibility for an environmental problem is most closely identified with the well established principle that the costs of reducing pollution "should be reflected in the cost of goods and services which cause pollution in production and/or consumption"[28] – the so-called *polluter pays principle* (PPP). This principle differs from the other two approaches, in that it can be viewed both as a condition for economic efficiency and as an ethical proposition. Under a strict "economic" interpretation, the principle requires that abatement effort be allocated so that the marginal cost of abatement and the marginal damages caused by emissions are equalized. While this requires a slight amendment in an international context – so that a country's marginal abatement cost is now set equal to its marginal attributable damages – the underlying principle remains intact. An alternative "ethical" interpretation of the principle would be that the distribution of cost burdens should reflect the relative levels of damages caused by each country's emissions.

However – as Sliggers and Klaassen (1994) observe – in an international context, the polluter pays principle may not be perceived as being fair, because the implied reductions in emissions will depend on exogenous factors such as meteorology, receptor sensitivity, and the geographic location of the countries. Hence, two countries with identical characteristics (population, wealth, etc.) and

---

[26] Those principles, attributes and outcomes which are important in co-ordinating agreement (Schelling 1960).
[27] There is an inherent inconsistency in this statement of principle in the context of transboundary pollutants, since it is not possible for all countries to simultaneously have the right to pollute and to be protected from pollution. It would be more meaningful to replace the "and" with "or", in which case the interpretation of the principle depends on the prevailing property rights regime.
[28] OECD Council, 14 November 1974 C(1974) 223, Paris 1974.

identical emissions may face different burdens purely because of an "accident" of geography.

The second approach emphasises the benefits accruing to a country as a result of the reduction in pollution – the so-called beneficiary pays principle. This approach is well established in relation to the provision of public goods (Young 1994), and has numerous applications in the area of natural resource management (e.g. fishing and hunting licences). However, in the case of anthropogenic pollution, this approach is more accurately described as the *victim pays principle* (VPP) since it is the actions of other countries which have (at least in part) created the environmental problem that is being addressed by the programme. Again, the magnitude of the damages in a particular country – and hence the magnitude of the environmental benefits it will receive from the abatement programme[29] – will be affected by its geographical location. For example, the climate may pre-determine the type of crops that can be grown in that country.

Countries also differ in terms of their ability to pay for any burdens attributed to them. At the interpersonal level, the *ability to pay principle* (APP) has long been recognised in relation to the raising of public finances (Young 1994). More recently, differences in ability to pay have also been reflected in the formulation of environmental policy at both the national and international level. In particular, Karadeloglou, *et al* (1995) identify a general principle in European Union policy decisions that cost burdens should be skewed towards those member states with relatively high income levels. While the ability to pay principle has an intuitive appeal, it can (at least in theory) produce anomalous outcomes. For example, it could result in a country being required to shoulder a large proportion of the total burden, even though it causes insignificant damage to other countries and receives minimal benefit from the abatement programme.

Thus, none of the three "focal" principles can be considered objectively superior; each having its own attractions, but also certain drawbacks. If there was a consensus over which of the three alternatives should apply in a particular policy situation then the issue would be relatively straightforward, and it would simply be a matter of identifying the relevant reference base for the calculations. However, Albin (1994) and Young and Wolf (1992) have argued that it is more appropriate to use a "balancing" approach, under which a number of different principles are combined. This view is given some empirical support by Atkinson *et al* (1999) in their econometric analysis of individual's preferences for alternative burden-sharing principles in relation to a local environmental problem in Portugal. They find that – rather than exhibiting lexicographic preferences for particular burden-sharing principles – individuals are willing to trade-off between alternative principles.

Unfortunately, the different equity principles are likely to lead to conflicting conclusions as to how the burdens should be shared (Burtraw and Toman 1992).

---

[29] Assuming that there is a common air quality standard across all countries.

9.3 Equity and Burden-Sharing

Consequently there is a need to develop an appropriate rule for *balancing competing principles*. A number of ad hoc rules have been proposed in the literature. For example, Smith *et al* (1993) combine indicators of ability to pay and responsibility for the problem, with equal weighting given to each. A less arbitrary approach to the assignment of relative weights is provided by the econometric analysis of Atkinson *et al* (1999). On the basis of their reported regression coefficients[30], it is possible to infer the following weightings for the three principles:

- Polluter pays   50 %
- Beneficiary pays   5 %
- Ability to pay   45 %

While the authors note that their results should be treated with caution, the relative values of the weights are not surprising. As one might expect, the greatest emphasis is placed on polluter pays and the ability to pay. However, it is interesting to note that beneficiaries are expected to make some (albeit small) contribution to the costs of abatement, even when others are responsible for causing the pollution.

9.3.2
Institutional Framework

The majority of work on equity and burden-sharing has been undertaken in relation to the problem of global warming, and the authors have invariably assumed that some form of international permit trading will provide the institutional mechanism for the implementation of their chosen principles.[31] When the underlying equity principle is defined in terms of "entitlements" it is straightforward (at least computationally) to determine the appropriate initial allocation of permits. For "outcome" based principles the problem is considerably more complex, as it is necessary to predict the volume – and price – of permit transactions in order to estimate the resultant net abatement costs. In this latter case, the burden-sharing transfers are implicitly defined by the initial allocation of permits. However, permit trading is not the only institutional mechanism that can be considered. Banuri *et al* (1996) acknowledge the potential for using direct financial transfers in the context of global warming, while Sliggers and Klaassen (1994) propose the setting up of an explicit "acidification fund" to share the financial burden of the UN/ECE Second Sulphur Protocol – into which well-off countries would contribute, and from which the less well-off countries would receive payments.

---

[30] The reported coefficients are for dummy variables relating to characteristics of different groups of people, e.g. whether they were responsible for the pollution, etc. As such, the coefficients represent the additional amount that a group with that characteristic should be expected to pay. The relative weights for the three principles reflect the relative values of the three coefficients.

[31] See, for example, Beckerman and Pasek (1995); Bohm and Larsen (1994); Kverndokk (1992); Rose (1992); Rose and Stevens (1993).

As has been discussed in *Sect. 9.2.2*, it is not feasible to implement European wide permit trading in the case of tropospheric ozone. Consequently, any system of compensating financial transfers will have to be implemented explicitly through the introduction of some form of *ozone fund*. A possible framework for such a fund is set out below. While it is more general in its structure than the acidification fund put forward by Sliggers and Klaassen (1994), offering greater flexibility in the detail and nuance of its implementation, it can accommodate their proposal as a special case. The framework is characterised by:

- the definition of those countries required to contribute to the fund, and those eligible to receive payments from the fund;
- the total size (or scale) of the fund;
- the rules for calculating contributions to the fund, and receipts from the fund.

In most cases, all countries would be expected to make contributions to the fund, and all would be eligible to receive payments from it. Of course, the values of the contributions and receipts would not be identical, and hence some countries would be net-contributors while others would be net-recipients. While the full participation of all countries in both sides of the fund has a certain political attraction, it would be perfectly possible to define sub-sets of countries as either contributors to, or recipients from, the fund. For example, under the Sliggers and Klaassen (1994) proposal, only those countries with above average GDP per capita are required to contribute to the fund, while only those countries with below average GDP per capita are eligible to receive payments.

The gross size of fund may be defined in absolute terms (i.e. a fixed amount of money), with the value either being set exogenously, or determined endogenously on the basis of the characteristics of the recipient countries. The latter approach is adopted by Sliggers and Klaassen (1994), who use a formula incorporating the abatement costs incurred by recipient countries and their relative per capita incomes to calculate the size of the fund. Alternatively, the scale of the fund may be defined relative to the total abatement costs of the recipient countries (which in most cases would be all countries). If the scale factor is set at a level less than 100% then burden-sharing is said to be "partial". There are a number of reasons why this may be desirable. Firstly, requiring countries to take sole responsibility for a proportion of their abatement costs is consistent with the original sentiments of the OECD polluter pays principle (see *Sect. 9.3.1*).[32] Secondly, the estimated abatement costs used in the calculations – which are usually based on technical measures only – are likely to exaggerate the actual costs that will be incurred. While the extent of this overestimation is difficult to predict, it can sometimes be considerable.[33] Consequently it may be prudent to set the scale factor at a level significantly lower than 100%.

---

[32] In this respect, the scale factor provides an alternative method for balancing the polluter pays principle with other principles such as ability to pay.

[33] The actual abatement costs under the US EPA Acid Rain Program were up to 33% lower than the forecast values (Ellerman *et al* 1997).

For a given burden-sharing principle (X = PPP, ... etc.), the net transfer to (or from) a country i is determined by the gross size of the fund, and by the difference between that country's share of total contributions ($\text{Share}_x^c(i)$) and its share of total receipts ($\text{Share}_x^r(i)$):

$$\text{Transfer}_x(i) = \left(\text{Share}_x^c(i) - \text{Share}_x^r(i)\right) \times (\text{Scale factor}) \times \left(\sum_{i \in I} \text{Cost}(i)\right)$$

It follows directly that if a country has a higher share of contributions than it does of receipts, then the net transfer is positive (i.e. it is a net contributor to the fund). Correspondingly, if the situation is reversed, then the net transfer is negative (i.e. the country is a net recipient from the fund). The sum of the net contributions (or net receipts) defines the net size of the fund. Thus while the gross size is of the fund is fixed, the net size depends on the choice of burden-sharing rule.[34]

The rules for calculating each country's share of the total payments into the fund will depend on the particular equity principle that has been adopted, and these are discussed in detail in the next section. Regarding the countries' shares of the total receipts from the fund, these will usually be set equal to their respective shares of the total abatement costs. Under this approach, the payment received by each country is proportional to the abatement cost that it incurs, and hence it is only the distribution of payments into the fund that affects the post-transfer allocation of costs. However this need not necessarily be so, and the shares can be adjusted to reflect some relevant characteristics of the recipient countries. For example, Sliggers and Klaassen (1994) adjust each recipient country's share of the fund to reflect their respective per capita incomes, with the poorest countries receiving a relatively higher share.

In theory the contributions to the fund, and the receipt of payments from the fund, could be administered through existing European Union financial mechanisms. For example, the existing contributions of member states to the EU Budget could be adjusted to reflect net contributions to, or receipts from, a "virtual" fund. Alternatively, the budgetary contributions of all member states could be increased to reflect their respective contributions, and the Cohesion Fund could provide the vehicle for making payments to countries. However, in practice it is likely to be preferable to set up an entirely new fund, where the transfers are kept separate from other financial flows within the community. Not only would this reduce the potential for the burden sharing issue to get subsumed in wider political and financial negotiations, it would also facilitate potential links with any burden sharing schemes that may be introduced as part of future UN/ECE Protocols.

---

[34] The distinction between gross size and net size is equivalent to the distinction between the value of the permits allocated and the value of permits traded in a permit based system.

9.3.3
Calculation of Transfer Payments

In this section, the general principles and approaches identified in *Sect. 9.3.1* are translated into specific operational formulations. Transfer payments are calculated for one particular INFOS reduction scenario – a 33 % gap closure for AOT60, with a 50 % weighting on acidification.[35] In each case, the gross size of the "ozone fund" is set equal to total EU-15 abatement costs; all countries are required to make contributions to the fund; and all countries are eligible to receive payments from the fund, with the distribution of receipts matching that of abatement costs, i.e.

$$\text{Share}_x^r(i) = \frac{\text{Cost}(i)}{\sum_{i \in I} \text{Cost}(i)} \quad \text{(where } x = \text{PPP, VPP, APP, BCP)}$$

The total EU-15 abatement costs under the policy scenario (compared to the 2010 trend scenario) are 39,500 million EURO_{95}, representing around 0.4 % of forecast EU-15 GDP in the year 2010. Around 80 % of total cost is borne by just five countries – France, Italy, Spain, Germany, and – to a lesser extent – the United Kingdom. However, the picture is very different when costs are measured relative to GDP. On this basis, it is Spain, Greece, Portugal, Ireland and Finland that bear the highest costs (approximately double the EU-15 average). As can be seen clearly from *Fig. 9.7*, the distribution of abatement costs between member states is highly regressive.[36]

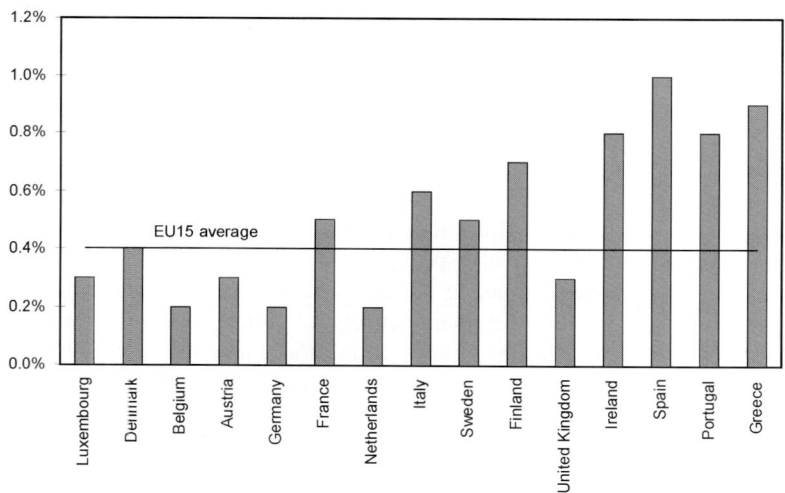

Fig. 9.7. Abatement costs as % of GDP

---

[35] See *Chap. 7* for a full description of the different INFOS reduction scenarios.
[36] In *Fig. 9.7* the sequence of countries (left to right) is based on their respective 1995 GDP per capita values (highest to lowest), measured under the purchasing power standard (PPS).

Polluter Pays Principle (PPP)

There are three issues which must be addressed when calculating burden-sharing transfers under the *polluter pays principle*. Firstly, it must be decided how the principle should be incorporated into an operational rule for calculating transfer payments. Secondly, each country's contribution to the aggregate damages must be determined. Finally, the value of the ozone related damages must be estimated for each country.[37]

As was noted in *Sect. 9.3.1*, one interpretation of the polluter pays principle is that the distribution of cost burdens should reflect the relative levels of damages caused by each country's emissions. There are several ways in which this interpretation could be put into practice. Costs could be apportioned on the basis of the relative contribution of each country to total damages prior to the introduction of the emissions reduction programme (i.e. pre-action), or on the relative contributions to the remaining damages after the programme has been implemented (i.e. post-action). Alternatively, the principle could be interpreted in terms of the relative total social costs attributable to each country under the optimal solution – i.e. their actual abatement costs plus their attributed damage costs. These alternative definitions are illustrated in *Fig. 9.8* for the special case of two countries with constant marginal attributed damages. Post-action damages attributed to each country ($i = 1,2$) are represented by D_i, gross abatement costs incurred by each country by A_i, and <u>net</u> benefits arising (to both countries) from each country's reduction in emissions by B_i.

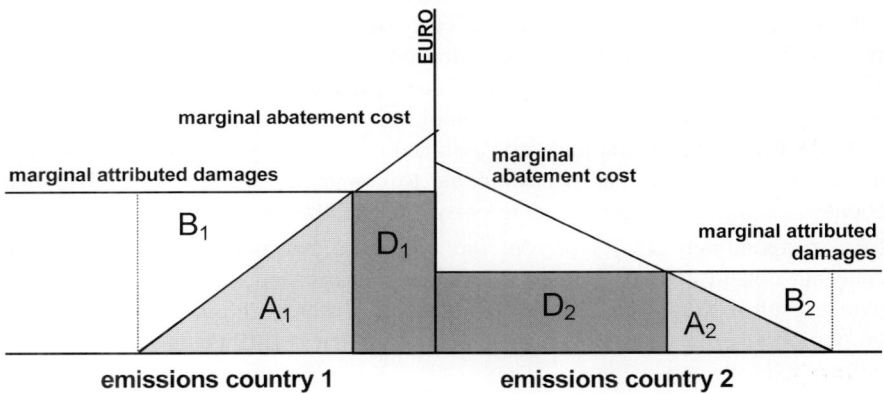

Fig. 9.8. Alternative interpretations of polluter pays principle

---

[37] The term "damages" relates to the monetary value (in $EURO_{1995}$) of the negative impacts of ozone on different receptors (people, crops, etc.). Similarly, the term "benefits" (i.e. reduction in damages) is expressed in monetary terms.

Under the first approach, total costs would be apportioned on the basis of the relative values of $A_i + B_i + D_i$, while under the second approach it is only the relative values of D_i that determine the allocation of costs. Of course, the two approaches will not in general give the same result, as can be seen in *Fig. 9.8* where country 1 faces a higher share of the total costs under the first approach, but a lower share under the second. Assuming that the proposed allocation of abatement efforts represents the true cost minimizing solution, then the third option (where costs are apportioned on the basis of the relative values of $A_i + D_i$) is equivalent to the relative costs that would arise if each country faced an optimal emissions tax equal to the marginal damages attributable to their respective emissions. Since this instrument is often advocated as a means of implementing the polluter pays principle, it would seem appropriate to use this interpretation as the basis for calculating transfer payments.[38] Consequently, the formula for calculating each country's share of contributions under the polluter pays principle is:

$$\text{Share}^c_{\text{PPP}}(i) = \frac{\text{Cost}(i) + \text{Attrib Damages}(i)}{\sum_{i \in I} (\text{Cost}(i) + \text{Attrib Damages}(i))}$$

For linear air pollution problems– such as with SO_2 – the second issue is straightforward, and the damages suffered by each country can be easily attributed back to source countries using the source-receptor matrix and the vector of country emissions. However, in the case of tropospheric ozone, where the relationship between ambient concentrations in one country and precursors emissions in another can be highly non-linear (particularly in the case of NO_x), the attribution of damages is much more complex.

In the following calculations, it is assumed that the proportion of ozone related damages in country j that is attributable to country i (i.e. β_{ji}) is equal to that country's contribution to the average ozone concentration in country j. Letting Z^N_{ji} (or correspondingly Z^V_{ji}) represent the difference between the ambient ozone concentration in country j under the policy scenario and the concentration that would result if NO_x (VOC) emissions in country i were to be set to zero, then the proportion of ozone related damages in country j that is attributable to country i is defined to be:

$$\beta_{ji} = \frac{Z^V_{ji} + Z^N_{ji}}{\sum_{i \in I}(Z^V_{ji} + Z^N_{ji})} \qquad Z^N_{ji} = z^N_j(e^N_i | e^N_{-i}, e^V) - z^N_j(0 | e^N_{-i}, e^V)$$

$$Z^V_{ji} = z^V_j(e^V_i | e^V_{-i}, e^N) - z^V_j(0 | e^V_{-i}, e^N)$$

---

[38] This does not mean that the allocation of abatement costs is the same as that which would occur under the "economic" interpretation of the polluter pays principle (see *Sect. 9.2.1*).

where $z_j^N(e_j^N \mid e_{-i}^N, e^V)$ is the "restricted" ozone function for country j with respect to NO_x emissions in country i, e_i^N is the NO_x emissions of country i under the policy scenario, and e_{-i}^N is the vector of NO_x emissions for all other countries (correspondingly for VOC).[39]

Ideally one would like to use the EMEP oxidant model (Simpson *et al* 1997) to calculate the ozone levels in each country when precursor emissions are set to zero in a particular country. Unfortunately the model is not designed to cope with such drastic shocks to the system, and the results of such an exercise would not be very meaningful. Alternatively it is possible to estimate the value of Z_{ji}^N (and correspondingly Z_{ji}^V) indirectly, utilizing the fact that the desired value is given by the integral:

$$\int_0^{e_i} z'(e_i \mid e_{-i}) \cdot d\,e_i$$

In order to determine the (approximate) value of this definite integral, it is necessary to estimate the values of $z'(\cdot)$ in the interval $[0, e_i]$. To this end, the EMEP model was used to calculate "source-receptor" matrices for 40 % reductions in emissions of NO_x and VOC (respectively) at two different levels of emissions – the *2010 trend scenario* (Base1), and a 60 % reduction on the trend scenario (Base2). In each case, the resultant slope coefficient (Δ) is equal to the slope of the restricted ozone function at some point on the closed interval $[0.6 \times BaseX, BaseX]$.[40] The exact point at which this holds true will depend on the particular shape of the restricted ozone function for that country / precursor. However, for simplicity it has been assumed that it occurs at the midpoint of the interval, and therefore that the calculated coefficients represent the slope of the function at $0.8 \times Base1$ and $0.32 \times Base1$ respectively (see *Fig. 9.9*), and that for:[41]

$$0 \le e_i \le 0.32 \qquad z'(e) = \Delta_2$$

$$0.32 \le e_i \le 0.8 \qquad z'(e) = \left(\frac{e_i - 0.32}{0.8 - 0.32}\right)\Delta_1 + \left(\frac{0.8 - e_i}{0.8 - 0.32}\right)\Delta_2$$

$$0.8 \le e_i \le 1 \qquad z'(e) = \Delta_1$$

This approximation allows for non-linearity of the restricted ozone function over the middle interval $[0.32, 0.8]$. In particular, in the case of NO_x, it allows the

---

[39] The "restricted" ozone function defines the impact on ozone concentrations in country j of changes in NO_x (VOC) emissions in country i when all other emissions are held constant.
[40] By the Mean Value Theorem (see Simon and Blume (1994), page 825), assuming that the function is continuous over the interval.
[41] For clarity, emissions have been normalised so that Base1 = 1. In this case, e_i can be interpreted as an emissions factor.

sign of the derivative to change over this interval. Thus, for example, if the value e_i lies in the middle interval then:

$$Z^N_{ji} = \quad + \int_{0.32}^{e_i} \left[\left(\frac{x - 0.32}{0.8 - 0.32} \right) \Delta_1 + \left(\frac{0.8 - x}{0.8 - 0.32} \right) \Delta_2 \right] \cdot dx \quad\quad 0.32\, \Delta_2$$

$$= \frac{1}{0.48} \left[(\Delta_1 - \Delta_2) \cdot \frac{x^2}{2} - (0.32 \cdot \Delta_1 - 0.8 \cdot \Delta_2) \cdot x \right]_{0.32}^{e_i} \quad + 0.32\, \Delta_2$$

Once the values of Z^N_{ij} and Z^V_{ij} have been estimated for all values of i and j, the coefficients of the "source-receptor matrix" $B = [\beta_{ji}]$ can be calculated directly. For this analysis separate matrices were calculated for AOT40 and AOT60 (the completed matrices are given in the appendix to this chapter), with the former being used for the attribution of crop damages, and the latter for the attribution of health damages.

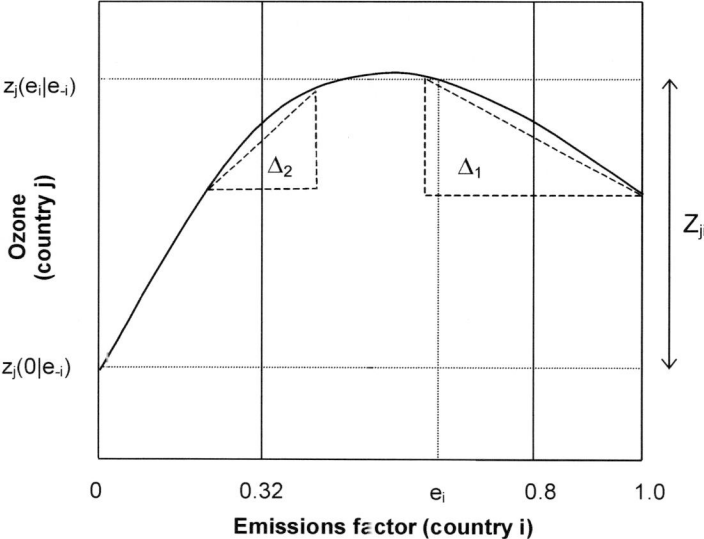

Fig. 9.9. Relationship between ozone concentration and NO$_x$ emissions

The final set of information required for the calculation of the transfer payments is the vector of estimated ozone related damages. These have been estimated using the integrated assessment model EcoSense, that has been developed as part of the EU funded ExternE project (Krewitt *et al* 1997).

9.3 Equity and Burden-Sharing

The model allows the valuation of ozone related impacts in relation to acute mortality effects, morbidity effects (asthma), and reductions in yields for ozone-sensitive crops (i.e. wheat, potatoes, rice, rye, oats, tobacco, and barley). The health mortality effects are calculated on the basis of a "Value of a Life Year Lost", while the valuation of morbidity effects assumes that WTP does not vary between countries.[42] *Table 9.6* (column 1) shows the calculated damages for each member state. While these values are only partial (they exclude chronic health effects, damages to rubber, or ecosystem impacts), they do include the most important impacts, and are likely to give a reasonable view of the distribution of damages between the countries.

Using the calculated source-receptor matrices, the actual damage values can be allocated back to the source countries, and aggregated to give the total attributable damages for each country. These values are shown in column 3 of *Table 9.6*, from which it can be seen that the distribution of attributable damages is broadly in-line with that of actual damages. The two main exceptions are Germany – where attributable damages are 7 % points higher than actual damages, and the United Kingdom – where attributable damages are 4 % points lower. The similarity between the two distributions is of course to be expected, given that – for many countries – their own emissions are the largest contributor to damages.

Table 9.6. Remaining damages under policy scenario in (mio. EURO$_{95}$)

Country	Actual Damages	Share	Attributed Damages	Share
Belgium	767	2.5%	1 317	4.3%
Denmark	398	1.3%	151	0.5%
Germany	6 286	20.7%	8 534	28.1%
Greece	1 807	6.0%	1 290	4.3%
Spain	2 825	9.3%	2 287	7.5%
France	4 964	16.4%	5 723	18.9%
Ireland	218	0.7%	26	0.1%
Italy	6 302	20.7%	5 970	19.7%
Luxembourg	26	0.1%	140	0.5%
Netherlands	1 048	3.5%	718	2.4%
Austria	623	2.1%	561	1.8%
Portugal	679	2.2%	959	3.2%
Finland	257	0.8%	13	0.0%
Sweden	511	1.7%	227	0.7%
United Kingdom	3 633	12.0%	2 427	8.0%
EU-15	**2755**	**100.0%**	**30 342**	**100.0%**

---

[42] See Krewitt *et al* (1999) for a discussion of both issues; in particular the two alternative approaches to valuing health mortality effects – "Value of a Life Year Lost" (VLYL) and "Value of Statistical Life" (VSL).

Victim Pays Principle (VPP)

Defining the cost apportionment rule under the *victim pays principle* is much more straightforward, with the formula for calculating each country's share of total contributions given by:

$$\text{Share}^c_{vpp}(i) = \frac{\text{Benefit}(i)}{\sum_{i \in I} \text{Benefit}(i)}$$

Again, the EcoSense model has been used to estimate the reduction in damages under the policy scenario, compared to the 2010 trend scenario. As can be seen from *Table 9.7*, the benefits arising from the reduction in emissions are highly concentrated, with four countries (Italy, France, Spain and Greece) accounting for over 80 % of the total; this being driven almost entirely by the increases in agricultural yields in those countries.

Ability to Pay Principle (APP)

Rose *et al* (1998) propose a general operational rule for the *ability to pay principle* that equalizes countries abatement costs as a proportion of GDP. This implies that each country's share of total contributions is equal to their share of GDP, i.e.

$$\text{Share}^c_{app}(i) = \frac{\text{GDP}(i)}{\sum_{i \in I} \text{GDP}(i)}$$

Unlike the resource bases used in the other two cases (which are assumed to remain constant in real terms), GDP will be growing over time. Since there is likely to be a divergence in the rates of growth between member states, the calculated share of contributions will change over time.

This can be seen in *Table 9.8*, which shows the contribution shares based on the 1995 actual GDP figures and the forecast values for 2010. For most countries it does not matter greatly which year is used for the reference base. However, for Germany, Italy and the United Kingdom, the differences are sufficiently large to affect the resultant values of their respective transfers by several hundred million $EURO_{95}$. In this analysis it has been implicitly assumed that the cost burden will not arise until the year 2010, and hence that burden-sharing transfers would not start until that time.[43] Consequently the calculations have been based on the forecast GDP figures for that year.

---

[43] In practice, of course, there would not be a step change in abatement in 2010, but abatement costs would rise steadily over the period 2000-2010.

9.3 Equity and Burden-Sharing

Table 9.7. Reduction in damages under the policy scenario in (mio. EURO$_{95}$)

Country	Health Benefits	Crop Benefits	Total Benefits	Share
Belgium	8.5	8.4	16.8	0.9%
Denmark	2.7	7.3	10.0	0.5%
Germany	62.1	90.1	152.2	7.7%
Greece	14.9	182.2	197.1	10.0%
Spain	93.2	275.8	369.0	18.8%
France	126.1	310.3	436.4	22.2%
Ireland	1.5	1.1	2.6	0.1%
Italy	104.5	540.7	645.2	32.8%
Luxembourg	0.6	0.4	1.1	0.1%
Netherlands	8.4	7.1	15.4	0.8%
Austria	8.6	7.1	15.7	0.8%
Portugal	12.9	30.9	43.8	2.2%
Finland	5.0	1.1	6.1	0.3%
Sweden	6.2	2.7	8.9	0.5%
United Kingdom	28.8	18.4	47.2	2.4%
EU-15	**483.8**	**1 483.6**	**1 967.4**	**100.0%**

Of course, this introduces the possibility of forecast error into the calculations. In practice however – since the burden-sharing transfers would not commence until 2010 – it would be possible to recalculate the values on the basis of actual information available at that time.

Table 9.8. Gross Domestic Product (in mrd. EURO$_{95}$)

Country	GDP 1995	% of total	Growth factor	GDP 2010	% of total
Belgium	209.3	3.2%	1.37	287.3	3.1%
Denmark	138.0	2.1%	1.40	193.8	2.1%
Germany	1 837.4	28.4%	1.50	2 754.4	29.3%
Greece	87.8	1.4%	1.49	130.5	1.4%
Spain	428.0	6.6%	1.50	640.7	6.8%
France	1 174.3	18.2%	1.46	1 716.6	18.3%
Ireland	50.2	0.8%	1.63	81.6	0.9%
Italy	832.0	12.9%	1.39	1 152.5	12.3%
Luxembourg	13.2	0.2%	1.37	18.0	0.2%
Netherlands	304.8	4.7%	1.42	431.4	4.6%
Austria	176.7	2.7%	1.44	254.0	2.7%
Portugal	80.7	1.2%	1.60	129.2	1.4%
Finland	96.3	1.5%	1.53	147.1	1.6%
Sweden	176.8	2.7%	1.40	247.9	2.6%
United Kingdom	859.7	13.3%	1.41	1 208.9	12.9%
EU-15	**6 465.2**	**100.0%**	**1.45**	**9 393.9**	**100.0%**

Under this approach, countries would agree the burden-sharing rule in advance (based on the forecast values), but leave the calculation of actual transfers until later.

Balancing Competing Principles (BCP)

Given the assumption that has been made regarding the distribution of receipts from the fund, the "balancing" of the different principles need only consider each country's share of the total contributions under the respective rules. Using the weighting factors derived from Atkinson *et al* (1999) (see *Sect. 9.3.1*), each country's share of contributions under the balanced approach is given by:

$$\text{Share}^c_{bcp}(i) = 0.50 \times \text{Share}^c_{ppp}(i) + 0.05 \times \text{Share}^c_{vpp}(i) + 0.45 \times \text{Share}^c_{app}(i)$$

9.4 Discussion

The net transfers calculated under the three different principles, and under the balanced approach, are shown in *Fig 9.10*. It is clear from this that the cost burdens arising under the "efficient" allocation of abatement efforts cannot be considered as being equitable under any of the four alternatives. Consequently, if there is a desire – or a need – for the costs to be shared on a more equitable basis, some re-distributive transfer payments will be required.

Table 9.9. Comparison of transfer payments under alternative principles

Comparison	Sign of transfer		Magnitude of transfer	
	Different	Same	Different	Similar
APP vs. PPP	4	11	8	3
VPP vs. PPP	9	6	6	0
VPP vs. APP	11	4	4	0

Comparing the three "pure" equity principles, there is a considerable variation in the *net* size of the burden-sharing fund – ranging from 3,200 million EURO$_{95}$ under the *polluter pays principle* to 9,700 million EURO$_{95}$ under the *victim pays principle*. In terms of the impact on individual countries, *Table 9.9* provides a breakdown of the results of pair-wise comparisons between the three alternatives. There is some degree of consistency between the transfers arising under the *polluter pays principle* and those arising under the *ability to pay principle* – the transfers have the same sign for eleven of fifteen the member states, and of these three have a similar magnitude. However, there little consistency between the transfers under the *victim pays principle* and the other two principles. Far fewer

9.4 Discussion

countries have the same sign of transfer, and in none of these cases are the transfers of a similar magnitude.

These results tend to confirm the findings of Burtraw and Toman (1992) that the adoption of different equity principles can lead to conflicting conclusions as to how the cost burdens of an international abatement programme should be shared. As such, they provide further justification for adopting a balanced approach to the calculation of burden-sharing transfers.

Under this approach, the net fund size is 5,200 million $EURO_{95}$. The main contributor to the fund is Germany (accounting for 62 % of net contributions), with lesser contributions from the United Kingdom (15 %), the Netherlands (10 %) and Belgium (8 %). The main beneficiaries from the fund are Spain (accounting for 47 % of net receipts), Italy (16 %) and France (12 %).

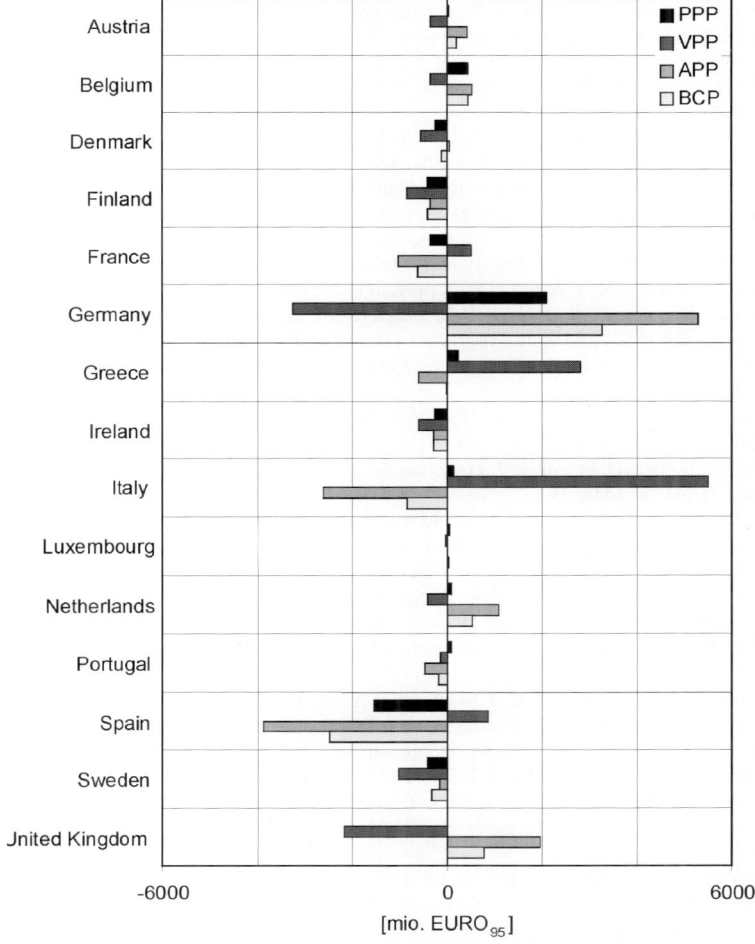

Fig. 9.10. Transfer payments under alternative equity principles in (mio. $EURO_{95}$)

9.5 References

Albin C (1994) Rethinking Justice and Fairness: The Case of Acid Rain Emissions Reductions. Review of International Studies, 21, pp 119-143

Atkinson G (1998) Equity, Burden Sharing and Pollution Abatement in Europe. CSERGE Working Paper, GEC 98-10

Atkinson G, Machado F and Mourato S (1999) Balancing Competing Principles of Environmental Equity. Paper presented at the 9th Annual Conference of the European Association of Environmental and Resource Economics, Oslo, June 25-27 1999

Banuri T, Goran-Maler K, Grubb M, Jacobson H K and Yamin F (1996) Equity and Social Considerations. in: IPPC, Climate Change 1995 – Economic and Social Dimensions of Climate Change. Contributions of Working Group III to the Second Assessment Report of the Intergovernmental Panel on Climate Change. Cambridge University Press, Cambridge

Barzel Y (1997) Economic Analysis of Property Rights. 2nd Edition, Cambridge University Press, Cambridge

Baumol W J and Oates W E (1971) The Use of Standards and Prices for Protection of the Environment. in: Swedish Journal of Economics, 73, pp 42-54

Beckerman W and Pasek J (1995) The Equitable International Allocation of Tradable Carbon Emission Permits. Global Environmental Change, 5(5), pp 405-413

Bohm P and Larsen B (1994) Fairness in a Tradable-Permit Treaty for Carbon Emissions Reductions in Europe and the former Soviet Union. in: Environmental and Resource Economics, 4, pp 219-239

Burtraw D and Toman M A (1992) Equity and International Agreements for CO_2 Containment. in: Journal of Energy Engineering, 118(2), pp 122-135

Ellerman A D, Schmalensee R, Joskow P L, Montero J P and Bailey E M (1997) Emissions Trading under the U.S. Acid Rain Program: Evaluation of Compliance Costs and Allowance Market Performance. Center for Energy and Environmental Policy Research, Massachusetts Institute of Technology

Flexner S B ed (1987) The Random House Dictionary of the English Language. 2nd ed (unabridged). Random House, New York

Hahn R W (1984) Market Power and Transferable property Rights. In: Quarterly Journal of Economics, 99(4), pp 753-765

Hahn R W (1986) Trade-offs in designing Markets with Multiple Objectives. in: Journal of Environmental Economics and Management, 13, pp 1-12

Karadeloglou P, Ikwue T and Skea J (1995) Environmental Policy in the European Union. in: Folmer H, Gabel H L and Opschoor H eds: Principles of Environmental and Resource Economics: A Guide for Students and Decision-makers. Edward Elgar, Cheltenham

Klaassen G (1996) Acid Rain and Environmental Degradation: The Economics of Emissions Trading. Edward Elgar, Cheltenham

Krewitt W, Mayerhofer P, Friedrich R, Trukenmuller A, Heck T, Gressmann A, Raptis F, Kaspar F, Sachau J, Rennings K, Diekmann J and Praetorius B (1997) ExternE – National Implementation in Germany. Final report prepared for the European Commission, DGXII, JOULE Programme, Contract JOS3-CT95-0010 EUR 18271

Krewitt W, Holland M, Trukenmuller A, Heck T, and Friedrich R (1999) Comparing Costs and Environmental Benefits of Strategies to Combat Acidification and Ozone in Europe. *forthcoming*

Krupnick A J, Oates W E and van den Verg E (1983) On Marketable Air Pollution Permits: The Case for a System of Pollution Offsets. in: Journal of Environmental Economics and Management, 10, pp 233-247

9.5 References

Kverndokk S (1995) Tradable CO2 Emissions Permits: Initial Distribution as a Justice Problem. in: Environmental Values, 4, pp 129-148
Luenberger D G (1965) Introduction to Linear and Nonlinear Programming. Addison Wesley, Reading, Mass.
Montogomery W D (1972) Markets in Licences and Efficient Pollution Control Programs. in: Journal of Economic Theory, 5, pp 395-418
Rose A (1992) Equity Considerations in Tradable Carbon Entitlements. in: Barrett S et al eds: Combating Global Warming UNCTAD, Geneva
Rose A and Stevens B (1993) The Efficiency and Equity of Marketable Permits for CO_2 Emissions. in: Resource and Energy Economics, 15, pp 117-146
Rose A, Stevens B, Edmonds J and Wise M (1998) International Equity and Differentiation in Global Warming Policy. in: Environmental and Resource Economics, 12, pp 25-51
Salmons R (1999) Iterative Zonal Permit Trading. CSERGE Working Paper, *forthcoming*.
Schelling T C (1960) The Strategy of Conflict. Harvard University Press, Cambridge, Mass.
Simon C P and Blume L (1994) Mathematics for Economists. Norton, New York
Simpson D and Eliassen A (1997) Control Strategies for Ozone and Acid Deposition – an Iterative Approach, Norwegian Meteorological Institute, EMEP MSC-W Note 5/97
Simpson D, Olendrzynski K, Semb A, Storen E and Unger S (1997) Photochemical Oxidant Modelling in Europe: Multi-annual Modelling and Source-receptor Relationships. Norwegian Meteorological Institute, EMEP MSC-W Report 3/97
Sliggers J and Klaassen G (1994) Cost Sharing for the Abatement of Acidification in Europe: The Missing Link in the New Sulphur Protocol. in: European Environment, 4(1), pp 5-11
Smith K R, Swisher J and Ahuja D R (1993) Who Pays to Solve the Problemand How Much?. in Hayes P and Smith K R (eds) The Global Greenhouse Regime: Who Pays? Earthscan, London
Stavins R N (1995) Transaction Costs and Tradable Permits. In: Journal of Environmental Economics and Management, 29, pp 133-148
Tietenberg T (1985) Emissions Trading: An Exercise in Reforming Pollution Policy. Resources for the Future, Washington, D.C.
Xepapadeus A (1997) Advanced Principles in Environmental Policy. Edward Elgar, Cheltenham
Young H P (1994) Equity: in Theory and Practice. Princeton University Press, Princeton
Young H P and Wolf A (1992) Global Warming Negotiations: Does Fairness Matter?. Brookings Review, Spring, pp 46-51
Zylicz T (1993) Improving Environment through Permit Trading: The Limits to a Market Approach. Beijer Discussion Paper, No. 23

9.6 Appendix

Source-Receptor-Matrix for AOT40

		\multicolumn{14}{c}{Source Country (in %)}
		AT	BE	DK	FI	FR	DE	GR	IE	IT	LU	NL	PO	ES	SE	UK
Receptor Country (in %)	AT	22.4	1.2	0.6	0.4	13.3	30.4	0.1	0.0	26.4	0.2	0.6	0.0	0.7	1.1	2.7
	BE	1.8	19.5	1.0	0.2	50.4	17.7	0.0	0.7	1.2	2.7	-9.0	0.0	2.7	1.4	9.5
	DK	0.4	3.0	23.4	1.4	5.8	30.1	0.0	0.7	0.2	0.1	3.4	0.0	0.1	13.4	17.9
	FI	0.0	0.8	7.8	25.9	1.4	11.7	0.0	0.0	0.0	0.0	1.8	0.0	0.0	34.4	16.1
	FR	0.7	4.3	0.3	0.0	57.0	15.3	0.0	0.3	5.6	0.6	1.7	0.2	8.3	0.3	5.5
	DE	4.2	4.5	2.1	0.5	21.9	51.6	0.0	0.3	3.5	1.0	0.5	0.0	0.9	2.1	7.0
	GR	0.9	0.2	0.2	0.1	2.0	4.3	58.0	0.0	22.0	0.0	0.4	0.0	0.6	0.3	1.1
	IE	0.2	3.7	1.1	0.0	15.3	16.9	0.0	13.9	0.2	0.0	4.6	0.0	1.4	0.8	42.0
	IT	2.3	0.7	0.1	0.1	14.8	5.8	0.7	0.0	70.4	0.0	0.6	0.1	2.4	0.3	1.6
	LU	1.3	10.4	0.4	0.1	42.6	28.8	0.0	0.2	1.4	5.2	1.9	0.0	1.6	0.5	5.6
	NL	4.3	24.6	4.7	0.7	56.9	5.1	0.0	2.4	0.9	1.5	-43.0	0.1	4.4	5.9	31.4
	PO	0.0	0.4	0.0	0.0	4.7	1.9	0.0	0.2	0.9	0.0	0.4	46.8	42.3	0.0	2.6
	ES	0.1	0.8	0.0	0.0	9.6	3.2	0.0	0.1	2.1	0.0	0.7	11.3	70.2	0.0	1.8
	SE	0.0	1.2	13.2	3.0	2.8	22.1	0.0	0.4	0.0	0.0	2.8	0.0	0.0	40.6	14.0
	UK	0.6	6.0	2.6	0.2	27.8	20.6	0.0	4.2	0.6	0.4	2.0	0.1	1.4	2.9	30.7

Source-Receptor-Matrix for AOT60

		\multicolumn{14}{c}{Source Country (in %)}
		AT	BE	DK	FI	FR	DE	GR	IE	IT	LU	NL	PO	ES	SE	UK
Receptor Country (in %)	AT	25.0	1.1	0.0	0.0	8.2	39.9	0.0	0.0	23.3	0.0	0.8	0.0	0.1	0.0	1.6
	BE	1.2	16.8	0.3	0.0	27.1	39.0	0.0	0.0	0.5	1.5	5.0	0.0	1.0	0.5	7.2
	DK	1.3	6.2	9.8	0.0	10.9	53.2	0.0	0.0	0.9	0.3	7.5	0.0	0.0	1.6	8.4
	FI	0.0	0.0	1.5	0.4	2.2	45.2	0.0	0.0	0.0	0.0	4.0	0.0	0.0	18.9	27.7
	FR	0.9	7.1	0.0	0.0	42.5	33.1	0.0	0.0	3.2	0.9	3.3	0.0	1.3	0.2	7.4
	DE	2.7	5.3	0.1	0.0	17.4	64.1	0.0	0.0	2.5	0.9	3.3	0.0	0.4	0.3	3.1
	GR	0.0	0.0	0.0	0.0	0.4	3.1	73.9	0.0	22.0	0.0	0.0	0.0	0.0	0.0	0.6
	IE	0.0	5.7	0.0	0.0	12.2	28.2	0.0	2.8	0.0	0.0	5.3	0.0	0.0	0.0	45.9
	IT	1.7	0.3	0.0	0.0	6.5	5.1	0.0	0.0	84.6	0.0	0.3	0.0	0.4	0.0	1.1
	LU	1.1	8.3	0.2	0.0	26.4	50.6	0.0	0.0	1.4	3.8	3.5	0.0	0.4	0.3	3.8
	NL	1.1	14.9	0.3	0.0	25.5	36.3	0.0	0.0	0.6	0.9	8.5	0.0	1.2	0.5	10.2
	PO	0.0	0.1	0.0	0.0	2.7	1.7	0.0	0.0	0.1	0.0	0.1	65.9	28.6	0.0	0.9
	ES	0.0	0.5	0.0	0.0	7.9	5.7	0.0	0.0	0.4	0.0	0.8	20.1	62.5	0.0	2.0
	SE	0.0	3.9	8.1	0.0	6.1	55.5	0.0	0.0	2.8	0.0	3.9	0.0	0.0	13.8	6.0
	UK	0.5	7.3	0.0	0.0	20.4	30.3	0.0	0.0	0.3	0.5	5.9	0.0	0.0	0.3	34.3

10 Macroeconomic Impacts of Abatement Strategies

Claudia Kemfert

10.1 Objective

Strategies to reduce ozone concentrations in Europe can have various economic effects. Each national economy, being confronted with the implementation of environmental protection measures, might react with different production, employment and competitiveness changes. Most environmental policy strategies, such as taxes or emissions permits, promote higher energy prices and thus increased costs for producers and consumers, stimulating economic substitution processes. As all European countries are linked by bilateral trade flows, national impacts have influences and spill-over effects on all other European and non-European countries.

Within the following chapter the impacts of different ozone abatement policies are evaluated with regard to macroeconomic parameters such as GDP, employment and competitiveness of each economy. After a brief description of the applied macroeconomic model, different abatement scenarios are compared with a *Business as Usual* (BAU) Scenario, i.e. taking no other additional environmental policies and regulations into account than those in place and in pipeline. One main emphasis lies on the comparison of scenarios including transfer payments derived from burden sharing rules with scenarios that do not incorporate any burden sharing.

10.2 Model Description

In order to investigate national or European environmental policies, applied general equilibrium models are especially useful. General equilibrium models determine the equilibrium prices due to an equalising of supply and demand of goods and products in each sector, while considering budget and production constraints over time. Substitution processes of input factors or consumption goods are mainly modelled by so-called CES production functions determining substitution opportunities due to substitution elasticities between factors and goods.

EAGE (European Applied General Equilibrium) is a recursive dynamic version of an applied general equilibrium model for the European economy including six European regions. It has been mainly designed to investigate European energy and environmental policy strategies like taxes on emissions or emission permits or generalised standards. A tradable permit scheme is included in EAGE. The regional dis-aggregation covers six key EU member countries accounting altogether for over 90 % of intra EU trade: Germany (GE), France (FR), United Kingdom (UK), Italy (IT), Spain (ES) and Denmark (DK). All other EU and non-EU regions are represented by an aggregate *Rest of the World* (ROW) importing from and exporting to the EU countries. Within the production behaviour of firms, five energy sectors are included (oil, gas, mineral oil, coal and electricity). Non-energy sectors are represented by their factor intensities, degree of factor substitution opportunities and price elasticities of demand. CES production or cost functions determine the substitution opportunities between capital, labour, energy and material (CLEM). The material aggregate is composed of non-energy inputs with fixed relations in all sectors. Within the energy aggregate substitution processes are determined by CES production or cost functions (see *Fig. 10.1*).

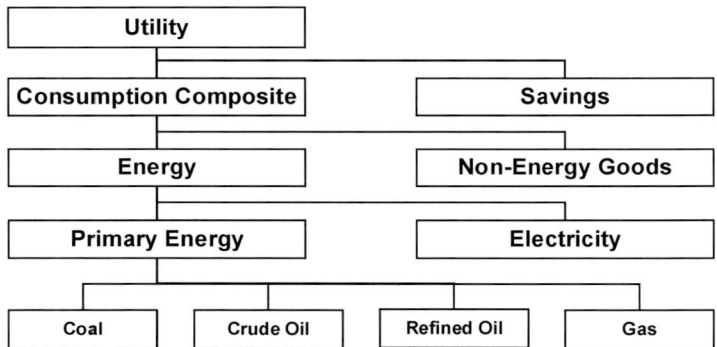

Fig. 10.1. Production Structure

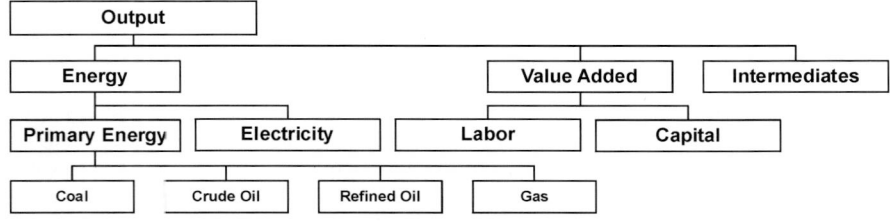

Fig. 10.2. Final demand Structure

The flexible CLEM specification of technologies allows energy input substitutions (fuel switch) and the substitution of energy by other inputs (energy saving). Consumption possibilities of households are expressed by CES demand or utility functions determining substitution opportunities between energy and non-energy goods (*Fig. 10.2*).

Primary factors are labour and capital. Due to the option of leisure, labour supply is elastic, which is accounted for by an empirically estimated labour supply elasticity. Capital stock is determined by investment and depreciation rates over time. The government produces public goods (like education, administrative services etc.) and transfers social contribution payments financed by tax policies. Domestic and imported goods and products are considered as homogenous imperfect substitutes, so called Armington goods (*see Fig. 10.3*).

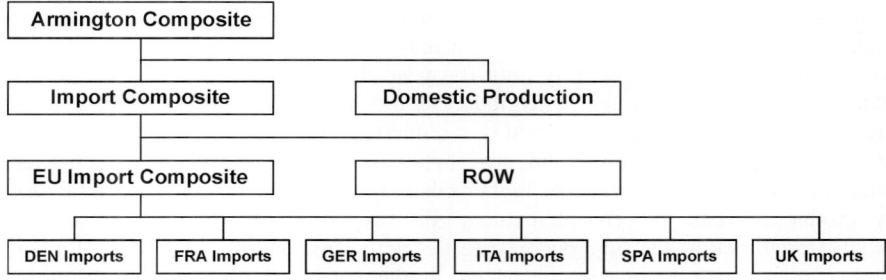

Fig. 10.3. Trade Structure in Each Sector

Exports and imports are determined by profit maximisation or cost minimisation of firms, resulting in an optimal tenacity of imports and exports. Foreign trade equilibrium is reached by a flexible exchange rate over a time horizon until 2010.

10.3
Model Calculations

10.3.1
Scenario Definition

In order to be able to compare all scenario results an equilibrium baseline is constructed according to assumptions about price and substitution possibilities and economic and emissions growth rates. Within this so called *Business as Usual* (BAU) scenario (cf. *Chap. 5*), the economies develop according to economic and environmental growth projections. The BAU scenario is used as a baseline reference to be compared with the scenarios including ozone abatement strategies.

Scenario 1 calculates the macroeconomic impacts of a 33 % Gap Closure of AOT 60, aiming at reducing ozone impacts on human health (see *Chap. 7*), without taking into account any transfer payments (sT: *sine transfer*). Scenario 2 investigates the same ozone reduction strategy for AOT 60, but includes transfer payments according to the burden sharing principle *ability to pay* (APP) (cf. *Chap. 9*).

Table 10.4. Scenario Definition

	Without transfer (sT)	With transfer (APP)
AOT 60 33 % Gap Closure	Scenario 1	Scenario 2

10.3.2
Model Results

Abatement costs due to the different reduction targets differ widely between European countries. In each scenario calculated annual abatement costs are imposed on the European countries Germany, Spain, France, Italy, the United Kingdom and Denmark annually until the year 2010. Ozone abatement measures impose additional costs and thus increase the price of capital resulting in a probable substitution of employment or a reduction of production investment. Due to production decreases, employment figures can decrease as well. Due to higher relative capital prices because of higher costs, labour and energy as input factors to production can be complementary or substitutionary in an ascertained production function causing decreasing or increasing output. In EAGE, higher capital prices induce on the one hand higher production costs (resulting in production decreases due to the costs structure of each firm).

On the other hand capital can be substituted by national or foreign energy or labour within the production function. Without transfer payments as a method of compensation, higher costs for each individual country triggers production losses finally resulting in employment losses. Including transfer payments, the economic burden of negatively affected countries can be reduced. National and international price changes cause substitution and adjustment processes because of linked trade relations. Thus, price increases – like a bilateral induced tax – cause effects depending on existing distortions of the economic system. All countries are linked interdependently causing highly complex trade spill over effects, which can hardly be separated.

Scenario 1

Fig. 10.4 displays the deviations of GDP from the baseline scenario for each of the investiegated European countries. Spain and France experience the highest GDP losses in absolute terms, while Germany and Italy face considerable decreases as well.

For Denmark and the United Kingdom effects on GDP appear to be marginal, which can be explained by the fact that both countries do not have to reduce their emissions by a large extent. While *Fig. 10.4* indicates mainly total abatement costs for France and Germany, in percentage terms of GDP Spain suffers highest GDP losses (see *Fig. 10.5*).

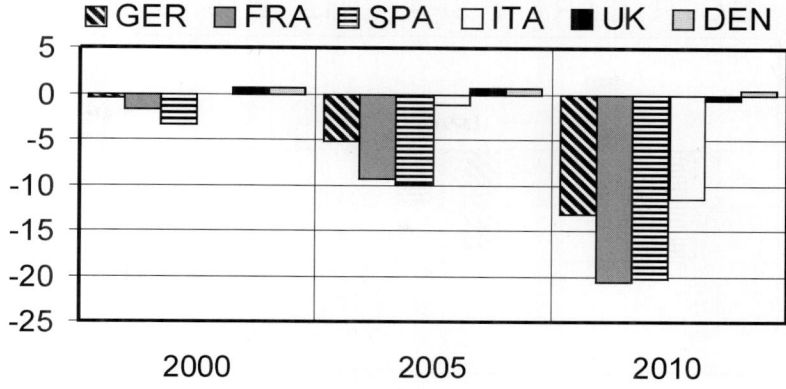

Fig 10.4. GDP Deviations in *Scenario 1* from BAU (in bill. ECU)

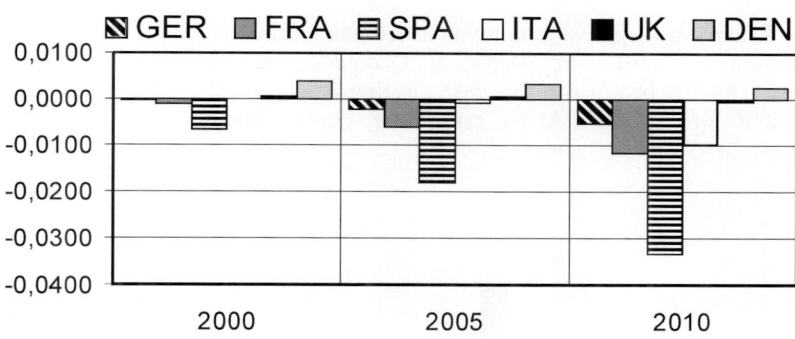

Fig. 10.5. GDP Deviations in *Scenario 1* from BAU (in % of GDP)

A comparatively rich country like Germany still has to pay a high amount of abatement costs, but in relative terms these losses affect the German economy far less than that of a comparably poorer country such as Spain (see *Fig. 10.7*). In order to compare the direct costs and payments all GDP figures are given as real values.

High abatement costs cause production losses inducing a reduction in employment for Spain, Italy, France, as well as Germany (*Fig. 10.6*). Since the European economy is a dynamic process linked by trade relations, welfare losses of countries with substantial trade relations induce production and welfare losses in all European countries.

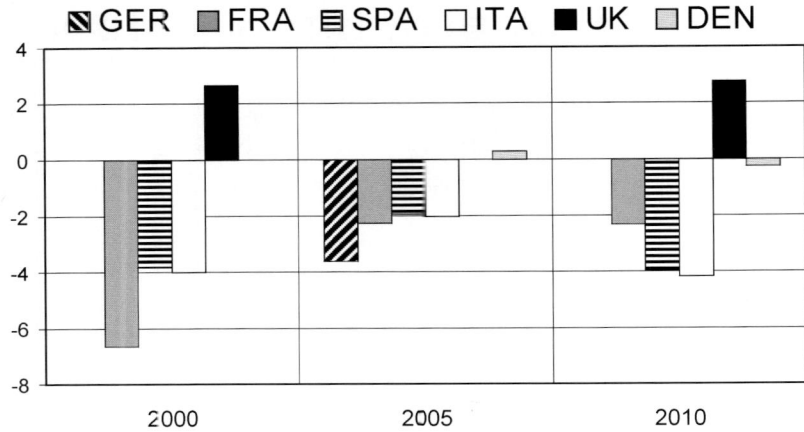

Fig. 10.6. Employment deviations of *Scenario 1* from BAU (in 1000 jobs)

For Denmark, negative employment effects in 2010 can be explained by substitution and spill over effects, as Denmark's additional costs are almost neglible. The UK benefits in this case ensuing employment increases by creating about 2000 jobs in 2010. At the same time, France, Spain and Italy experience negative employment effects.

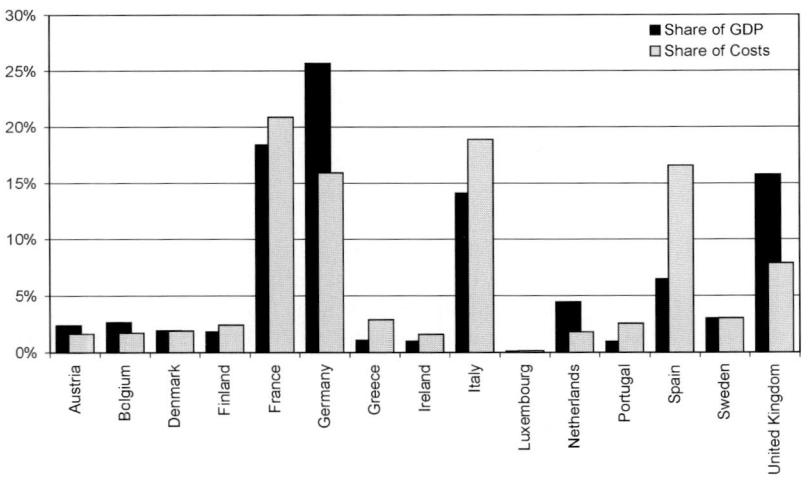

Fig. 10.7. Comparing shares of total EU15 GDP and shares of total abatement costs in 2010

Scenario 2

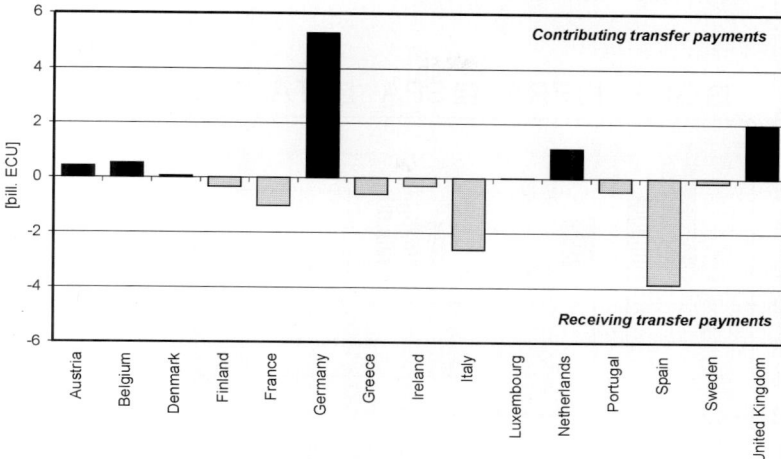

Fig. 10.8. Countries contributing and receiving transfer payments according to their ability to pay (APP)

Scenario 2 incorporates transfer payments according to the *ability to pay* principle (APP). *Fig. 10.8.* shows, which countries have to contribute, respectively receive transfer payments.

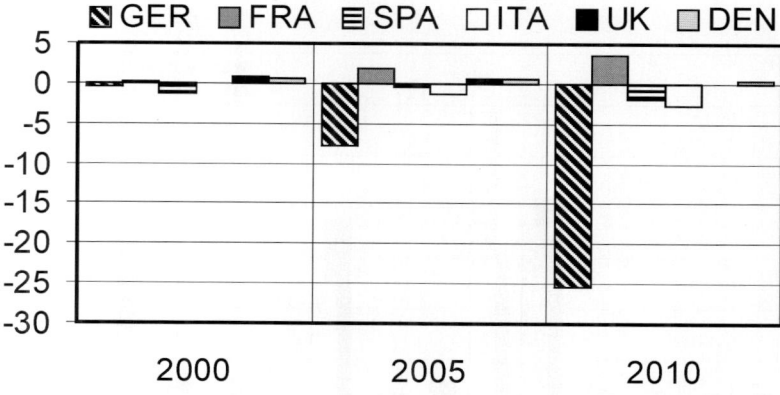

Fig. 10.9. GDP deviations of *Scenario 2* from BAU (in bill. EURO per year)

Germany, as a major transfer payer within this burden sharing scheme (see *Chap. 9*), suffers most by high abatement costs and additional transfer payments (*Fig. 10.9*), leading to employment losses of more than 10 000 jobs after 2005.

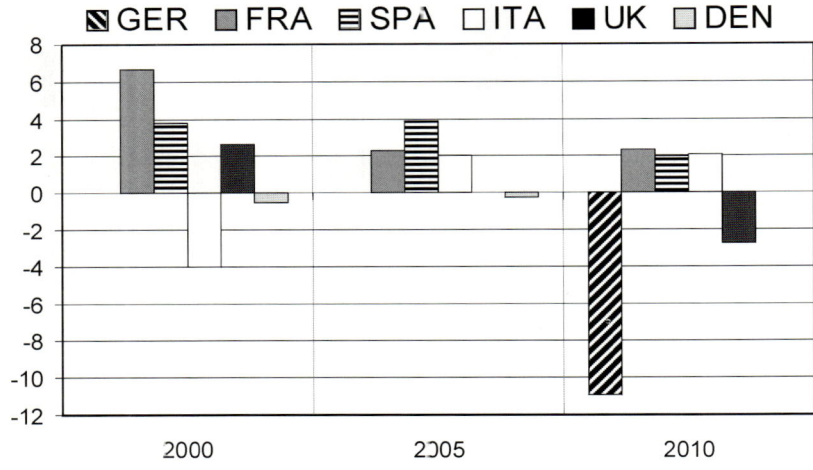

Fig. 10.10. Employment deviations of *Scenario 2* from BAU (in 1000 jobs)

Countries receiving transfer payments such as France, Spain and Italy gain by production and employment increases (see *Fig. 10.10*), but especially Italy and Spain, being import-oriented, are negatively affected by Germany's loss in GDP.

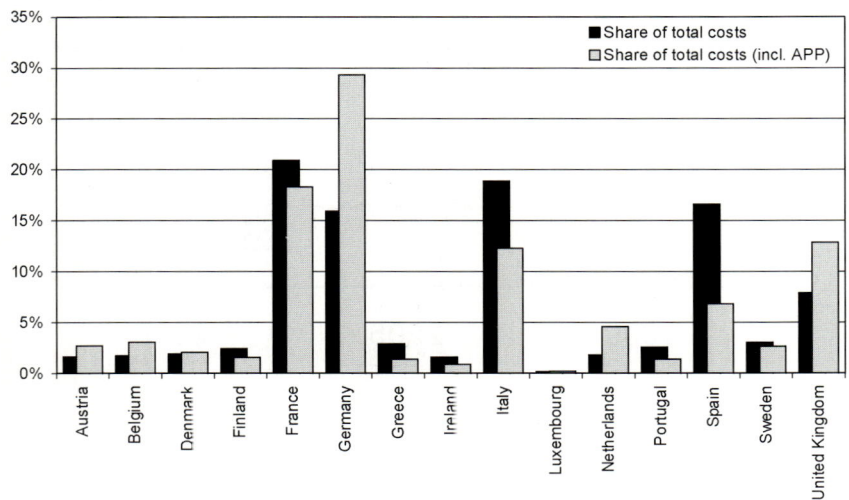

Fig. 10.11. Comparing shares of total costs in 2010 without and with transfer payments included

The UK and Denmark suffer primarily from spill over effects due to price and cost changes and negative effects to welfare, i.e. GDP losses in Germany. Italy encounters slight GDP reductions, but can finally benefit by an increase of employment in 2010. This is mainly resulting from transfer payments and the distributional effects of burden sharing (*Fig. 10.11*).

10.4
Effects on Import and Export

The impacts on the competitiveness of all European countries – assessed by import and export development – varies between regions and scenarios (*Figs. 10.12* and *10.13*).

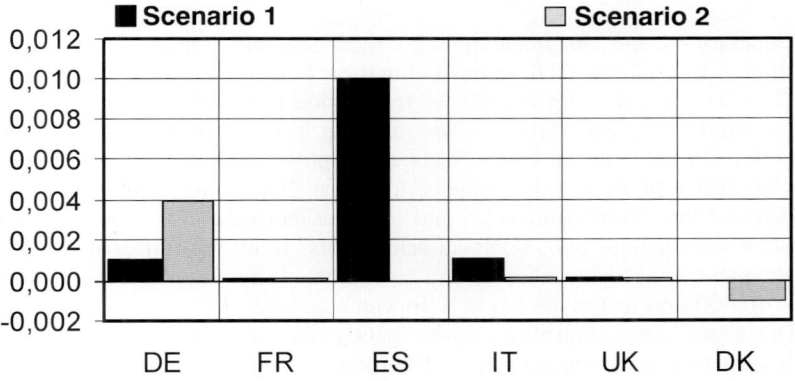

Fig. 10.12. Import changes in percent of GDP

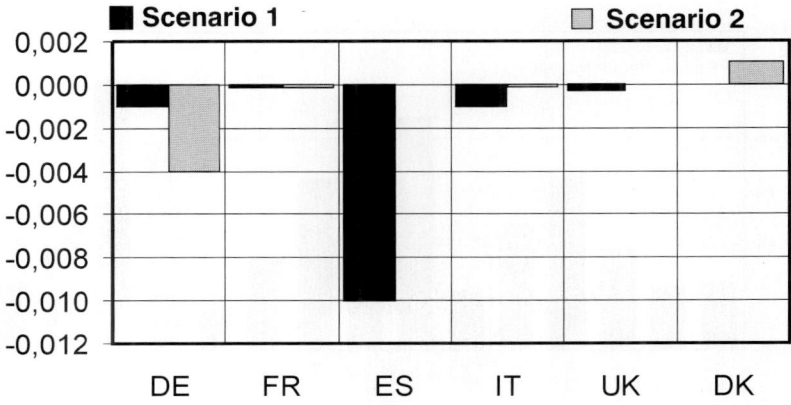

Fig. 10.13. Export changes in percent of GDP

For Germany, the introduction of transfer payments (*Scenario 2*) means considerably higher total costs (abatement cost plus transfer payments) and thus the effects on imports and exports are higher than in *Scenario 1*. Spain clearly benefits from the distributional effects of applying a burden sharing princple, as the severe impacts of an abatement strategy without burden sharing are equalized. This can be explained by the type of burden sharing rule used, since the *ability to pay* (APP) takes the relative economic strength of each country into account. For all other countries, import and export effects are almost negligible under *Scenario 2*, showing the importance of burden sharing especially for the – comparatively – economically weaker countries.

10.5
Summary and Outlook

In summary, ozone abatement policies induce higher total and marginal costs leading to decreases of GDP in most countries. For the scenarios considered here, GDP losses amount to up to 0.03 %. In addition to that, higher abatement costs induce production and welfare losses resulting in substitution processes between input factors. Macroeconomic effects are highly complex and interdependent, causing different economic impacts for each European country investigated. Because of the interdependencies and relations between all European countries, negative production and GDP developments result in different effects on employment. Employment losses can be estimated to a scale of several thousand up to 10 000 jobs lost in the larger European countries.

The main conclusion from investigating the economic impacts of ozone abatement policies is that equity and burden sharing issues do play a vital role. Under equity aspects, the burden can be shared by a compensation method easing the burden of the most negatively affected countries, e.g. Spain and Italy. Under the burden sharing principle *ability to pay* (APP) all countries carry the same share of costs relative to their GDP, as indicated by *Fig. 10.14*.

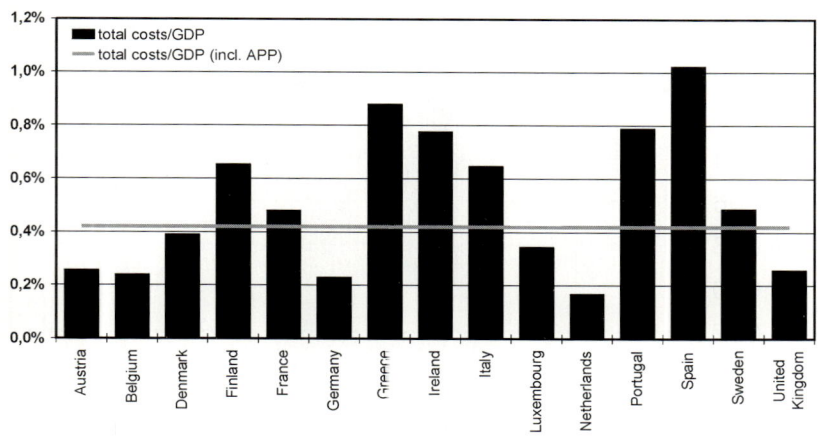

Fig. 10.14. Comparing cost burdens relative to GDP with and without burden sharing by APP

Within this investigation, most European countries are taken into account, covering about 90 % of European trade and more than 80 % of total EU15 GDP. However, important bilateral trade relations between Northern and Southern European countries are not reflected, probably resulting in an underestimation of potential effects. On the other hand, innovations resulting in more efficient abatement measures and additional opportunities to produce and export new abatement technologies are not taken into account in this analysis, which could reduce the negative macroeconomic effects.

10.6 References

Böhringer C, Harrison G and Rutherford T(1997) Sharing the Burden of Carbon Abatement in the European Union. confernce proceedings IEW / EMF

Böhringer C, Jensen J and Rutherford T (1997) The Costs of Carbon Abatement in Six EU Countries: Implications of Alternative Baseline Energy Projections

Kemfert C (1998) Estimated substitution elasticities of a nested CES production function for Germany. in: *Energy Economics* 20, pp 249–264

Kemfert C, Welsch H (1998) Energy-Capital-Labor Substitution and the Economic Effects of CO_2 Abatement: Evidence for Germany. in: *Journal of Policy Modelling*, forthcoming

Kemfert C, Pfaffenberger W. (1998) CO_2 taxation and competitiveness of the German Economy – An analysis with the Macroeconomic Information System (MIS). in: *International Journal of Global Energy Issues*, IEW / JSER (eds.) Vol 11 No. 3 / 4, pp 213–219

11 Summary and Conclusions

Rainer Friedrich and Stefan Reis

Summary

The high concentrations of tropospheric ozone all over Europe during summer episodes still is an unsolved problem of air pollution. In the years 1994 and 1995, the EU threshold for human health, 110 µg/m^3 (over an 8 h mean), was substantially exceeded, exposing a large share of the European citizens to increased ozone levels, some even on more than 25 days during the year.

The European Commission issued the Directive on Air Pollution by Ozone (92/72/EEC) in 1992, setting ozone thresholds and air quality targets in order to improve ambient air quality in the EU. Furthermore, measures to reduce precursor emissions have been implemented or will be implemented initiated by the EC Large Combustion Plants (LCP) Directive, the EURO I – IV emission standards for road transport vehicles, the EC Solvent Directive, the UNECE Convention on Long Range Transboundary Air Pollution and national activities. So, the first question to be answered within this study was, whether these activities and all related legislation and regulations in place and in pipeline will suffice to reduce precursor emissions to an extent that will reduce ambient concentrations of ground-level ozone below values that are supposed to be harmful for human health and plants. The relevant levels are: for human health an average value of 60 ppb (120 µg/m^3) during any 8 hour period, not to be exceeded; for crops the AOT40 (accumulated exposure over 40 ppb) value should not exceed 3 ppm.h accumulated from May to July and for forests AOT40 should not exceed 10 ppm.h from April to September.

The year 2010 has been chosen as time horizon to generate a base case or trend scenario, because it is close enough to allow projections of future activity levels with sufficient accuracy, but it allows sufficient time for market penetration of abatement measures. Results generated within this study show that the overall emission reduction in a business-as-usual scenario is projected to be 30 % for NO$_x$ and 37 % for anthropogenic NMVOCs of total EU15 emissions in 2010 compared to the 1990 level. (see *Figs. 11.1* and *11.2*)

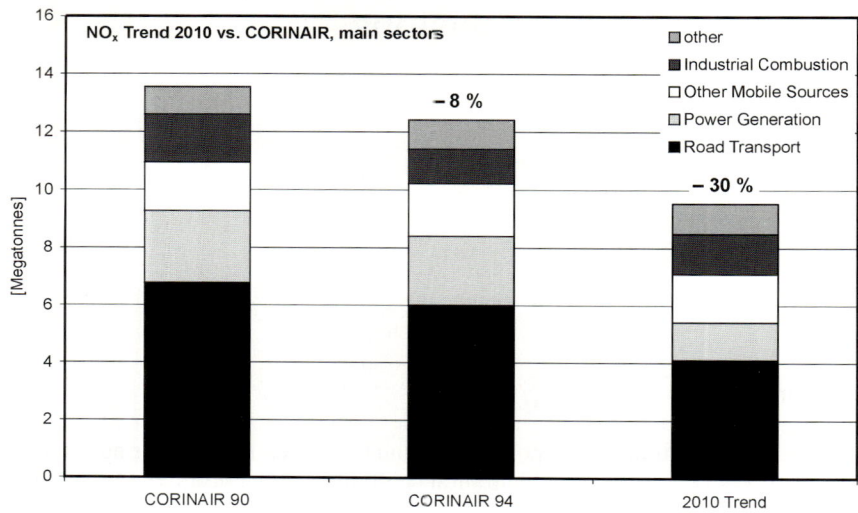

Fig. 11.1. Anthropogenic NO$_x$ emissions in EU countries – inventories and trend scenario 2010

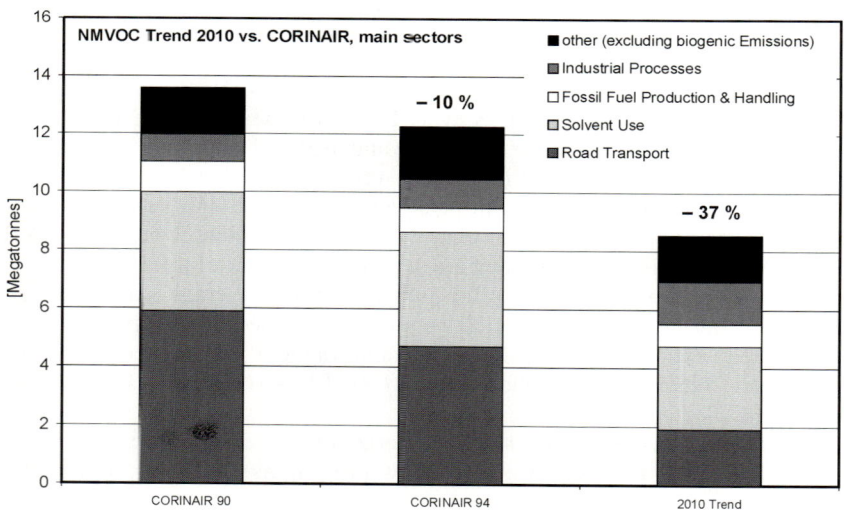

Fig. 11.2. Anthropogenic NMVOC emissions in the EU15 – inventories and trend scenario 2010

In spite of these significant reductions of ozone precursors, results from photo-oxidant modelling calculations still suggest exceedances of the ozone limit values mentioned above, on both the European level and for urban areas. The exceedance of thresholds is reduced considerably in the base case scenario, but especially the

60 ppb limit is still exceeded notably; on average over all EU land areas, the percentage of time, for which ozone levels above 60 ppb are calculated, is reduced from 4.7 % 1990 to 2.8 % 2010 (*Table 11.1*).

Table 11.1. Percentage of time for which ozone levels of 60-120 ppb are exceeded in the EU, as an average across all EU land-areas. Results are taken from the 5-year run of the EMEP model for the scenarios listed.

Ozone level (ppb)	Base 1990	Trend 2010
60	4.7	2.8
70	1.9	0.9
80	0.8	0.3
90	0.4	0.09
100	0.2	0.03
110	0.06	0.01
120	0.03	0.003

The (modelled) maximum concentrations of ozone in and around urban areas are significantly reduced in the trend scenario as well, but limit values are still exceeded. For example in and around Athens the calculated maximum 8h average ozone concentration is reduced from 165 ppb in 1990 to 128 ppb in the year 2010; whereas for 1990 exceedances of the 60 ppb threshold are calculated to occur during maximal 81 days, this is reduced to 76 days in 2010 (*Table 11.2*). When implementing a 33 % gap closure of AOT 60 (*Table 11.3*), exceedances of 120 ppb will become very rare, respectively are practically avoided all over Europe.

Table 11.2. Comparison of modelled ozone concentrations in the base year 1990 and for the trend scenario 2010

values in ($\mu g/m^3$)	Athens			Milan			Stuttgart		
Base Case 1990	Domain	Suburban	Urban	Domain	Suburban	Urban	Domain	Suburban	Urban
max. 8h average	331	273	248	162	149	127	306	283	267
max. 1h average	571	570	476	203	196	162	349	329	305
average	81	59	35	60	57	46	102	96	86
Trend 2010									
max. 8h average	257	257	230	154	146	127	213	197	195
max. 1h average	409	409	400	190	190	164	241	227	228
average	78	61	45	58	57	49	87	84	77

Thus, while the situation of high ambient concentrations of ozone will be improved to some extent in the year 2010, additional action has to be taken, if limit values are not to be exceeded. And since taking additional measures will impose additional costs on all EU countries, it is vital to find an optimal, i.e. least-cost, strategy to be implemented and assess its distributional and macroeconomic impacts.

Within this study, detailed abatement cost curves were calculated, taking into account relevant abatement measures for NO_x and NMVOCs that are currently available, or in a stage of market introduction, so that cost estimates could be obtained and the measures can be implemented by 2010. Of course, only measures in addition to those that are already taken into account in the base case scenario for 2010 are considered. The cost curves give the cost of reducing each unit of precursor emissions by ranking abatement measures according to their unit costs [cost/tonne of pollutant abated], and assessing the abatement potential for each source sector the respective measure could be applied to. In some cases, huge differences between countries arise from varying implementation degrees in the base year or differing sectoral structures (e.g. type of fuel used for electricity generation). Thus, the maximum reduction achievable by implementing all measures listed in the country specific cost curves as well as the total costs show a considerable variation. The possible additional emission reductions in 2010 on top of the base case in 2010 vary between 30 % and 50 % for NO_x and 10 to 25 % for anthropogenic NMVOC. Regarding total European emissions, a maximum reduction on top of the trend scenario of 41 % (NO_x) and 17 % in the case of anthropogenic NMVOC is possible, if technical measures currently available, but no improvements or new developments of technical measures or behavioural changes are taken into account. Mean unit costs for the reduction differ significantly between NO_x (2.6 kEURO/tonne pollutant abated) and NMVOC (21.6 kEURO/tonne pollutant abated), leading to total maximum costs of abatement of 10 bill. EURO (NO_x) and 33 bill. EURO (NMVOC) at the maximum reduction described above.

The approach to find an optimal strategy for emission reduction on top of the 2010 trend scenario included using an iterative optimisation model to relate emission reduction and costs to changes in the concentration of ground-level ozone. This model, named OMEGA, uses source-receptor-relationships calculated by the EMEP model to derive ozone concentrations for each grid cell at each iteration step, selecting the least-cost option for every next unit of emission reduction from the country specific cost curves. The calculations, however, made clear that even with all the additional measures identified and included in the cost curves, the above mentioned limits are still not reached.

So, as a target for optimisation, a gap closure of X % towards three different ozone limit values ($AOT40_{crops}$, $AOT40_{forests}$ or $AOT60_{health}$) was attempted, i.e. the 'gap' between the actual value and the threshold should on average be reduced by X %. With the established cost curves an optimised gap closure of up to about 33 % for AOT60 (health) and $AOT40_{forests}$ and up to about 15% for $AOT40_{crops}$ can be achieved. The necessary emission reductions to attain these gap closures are indicated in *Table 11.3*.

Table 11.3. Emission reduction (relative to 1990) necessary to achieve given Gap Closures (GC) of AOTxx

Total EU15	Optimisation scenarios		
	AOT60 33% GC	AOT40$_{crops}$ 15% GC	AOT40$_{forests}$ 33% GC
NO$_x$	- 47 %	- 45 %	- 42 %
NMVOC	- 47 %	- 41 %	- 39 %

The costs for these emission reductions amount to a range of 9 to 39 billion EURO (see *Table 11.4*). An analysis of different optional scenarios for the 60 ppb limit (with regard to human health impacts) reveals, that costs increase non-linear with increasing gap closures. A 10 % gap closure for AOT 60 for example causes costs of 1.4 billion EURO, a gap closure of 25 % costs about 11 billion, a gap closure of 33 % finally leads to costs of 39 billion EURO.

Table 11.4. Total abatement costs for the selected scenarios for the EU15

	Total Costs EU$_{15}$ (in bill. EURO)
AOT60 33% Gap Closure	39
AOT40$_{crops}$ 15% Gap Closure	17
AOT40$_{forests}$ 33% Gap Closure	9

As this is a considerable amount of money, the question arises, which impacts the implementation of these strategies would have on the economy, especially on economic growth and employment. To analyse these impacts, the general equilibrium model EAGE was used. Results for the 33 % gap closure for human health show a decline of GDP for most countries. Highest relative reductions of GDP are calculated for Spain being – 0.03 % of GDP. However, for instance Denmark sees a small increase in GDP (+ 0.002%) due to trade spill over effects. Employment effects also are in general negative, e. g. the model calculates a decrease in employment of ca. – 0.0001 % to – 0.0002 % in France, Germany, Spain and Italy. It is important to note, though, that these figures do not include positive impacts due to innovation e.g. by improving emission reduction measures and possible positive effects due to an increased exports of innovative products for environmental protection. The estimated reduction of GDP leads to an additional reduction of emissions of ozone precursors and so of ozone concentrations. However, this effect is so small, that it can be neglected.

As the EU draft Ozone Daughter Directive defines a limit value for ground-level ozone of 120 µg/m^3 (8h mean, cf. *Chap. 2*), with 20 – 25 exceedances being allowed, the scenarios investigated have been assessed according to their ability to achieve this limit. *Fig. 11.3* shows the exceedances of 60 ppb (at 20 °C and 1013 mb pressure, 1 ppb equals 2.00 µg/m^3) in the base case 1990 and the trend scenario 2010, indicating a significant improvement for most of Europe in the trend scenario already.

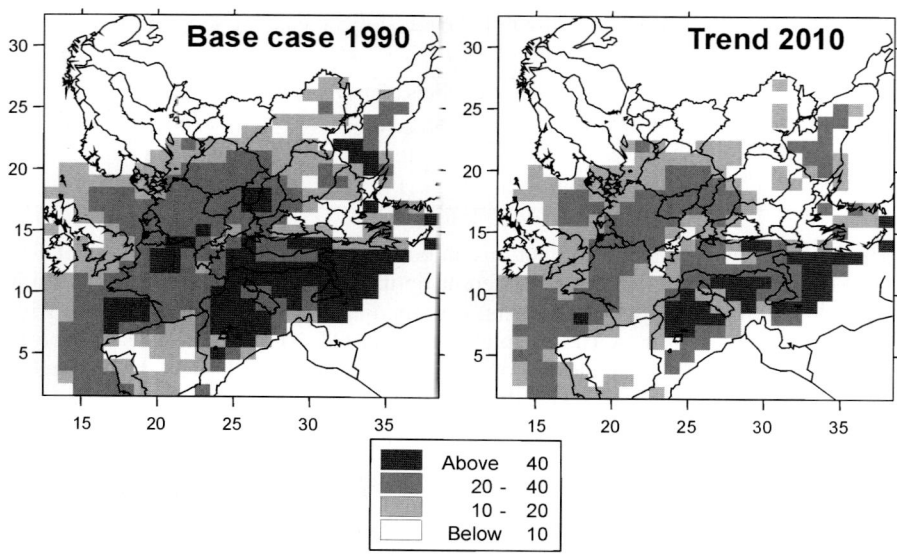

Fig. 11.3. Days of exceedance of 60 ppb in the base case 1990 and the trend 2010 scenario

In order to illustrate how the model predictions of exceedances vary over selected scenarios, the number of days in each grid square for which either one of the 6 hourly ozone values at 12 or 18 GMT exceeds 60 ppb has been calculated. The 5-year average results of this calculation are presented in *Table 11.5.*, indicating that on average the number of days exceeding 60 ppb is reduced from 19 in the base case 1990 to 13 in the trend scenario for 2010. A further reduction to almost 9 days is calculated for the optimised control scenarios. These figures are averaged over the whole EU land area.

The percentage of the EU land area in which the number of exceedance days is greater than 20 has been calculated as well. In 1990 over 46 % of the EU land area are predicted to have more than 20 exceedances of 60 ppb per year.

Table 11.5. Number of days where 60 ppb is exceeded (as an average over the EU area), and percentage of EU-area with more than 20 days of exceedance of 60 ppb[1] (120 µg/m^3) per year.

	Base 1990	Trend 2010	AOT40$_{crops}$ 15 % GC	AOT40$_{forests}$ 33 % GC	AOT60 33% GC
Number of days > 60 ppb	19.8	13.0	9.88	9.19	9.18
EU land area with > 20 exceedance days (%)	46.4	27.2	17.2	14.5	13.9

[1] at 20 °C and 1013 mb pressure, 1 ppb equals 2.00 µg/m^3

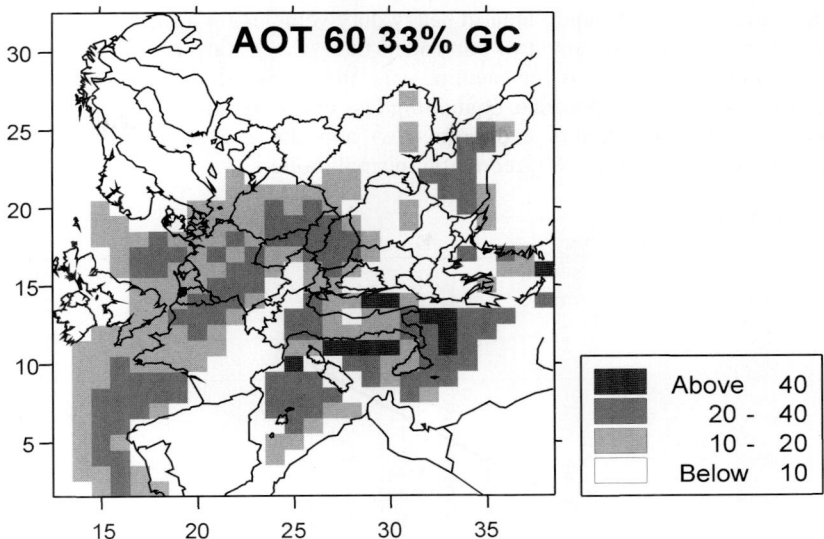

Fig. 11.4. Days of exceedance of 60 ppb in the optimised reduction scenario AOT60 at 33 % GC

This is reduced significantly to 27 % in the Trend 2010 case and further to below 14 % in the control scenarios (see *Fig. 11.4*), with exceedances remaining mostly in western France, Belgium and Germany, as well as northern Italy. The high number of exceedances over sea areas is expected, and is simply the result of the lack of dry deposition of ozone to sea surfaces. Without deposition, concentrations at ground-level (~ 1m) over sea areas are just as high as in the rest of the boundary layer. In contrast, over land-areas where deposition is important, ozone concentrations at ground level are substantially lower than the boundary layer average.

Analysing gap closures of up to 33 % does not mean that further reductions of exceedances of the ozone limits would not be possible; but, this will involve additional measures and technologies, that currently are not yet available on the market (so that they could not be included in the cost curves). Current and planned legislation is already quite stringent, for instance with regard to exhaust emission of new cars or new power plants and the use of solvents.

Furthermore, the industry branches which produce emission control equipment tend to concentrate more on the development of techniques that fulfil legal requirements more efficiently than on the development of techniques that reduce emissions further than required by regulation. So, additional possibilities to reduce emissions using techniques already available in the market especially include the application of these techniques to additional emission sources (smaller plants or applications, retrofitting and so on) or a speed up of market penetration. Nevertheless, a number of techniques are in development or are a topic of current research, that have the potential of a substantial further reduction of emissions. Techniques with considerable potential beyond the currently planned regulation include the use of SCR equipment in heavy duty vehicles for a further reduction of NO_x, the further improvement of controlled three way catalysts and of devices to reduce cold start emissions for gasoline cars, the extended substitution of solvent containing products by products containing less or no organic solvents (e.g. water-based or high solid paints and varnishes) and the use of primary measures (e.g. low NO_x burner) for NO_x reduction in small boilers.

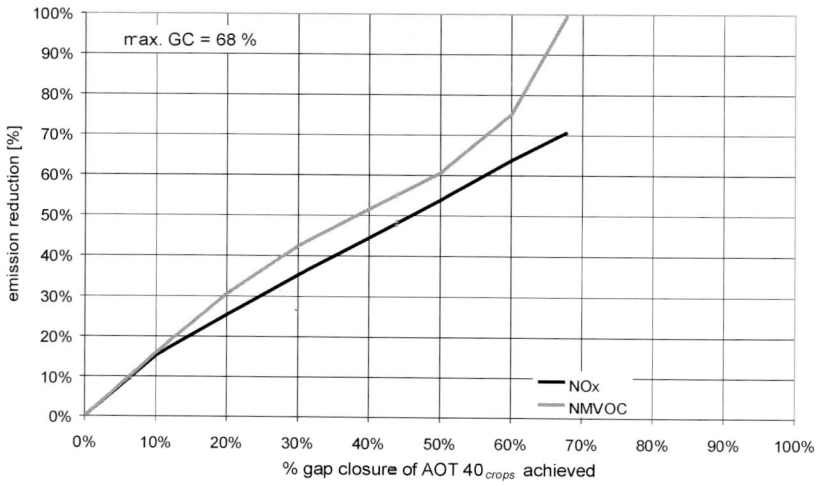

Fig. 11.5. EU total reductions of NO_x and anthropogenic NMVOC to achieve more stringent gap closures of $AOT40_{crops}$; reductions in % on top of trend scenario 2010

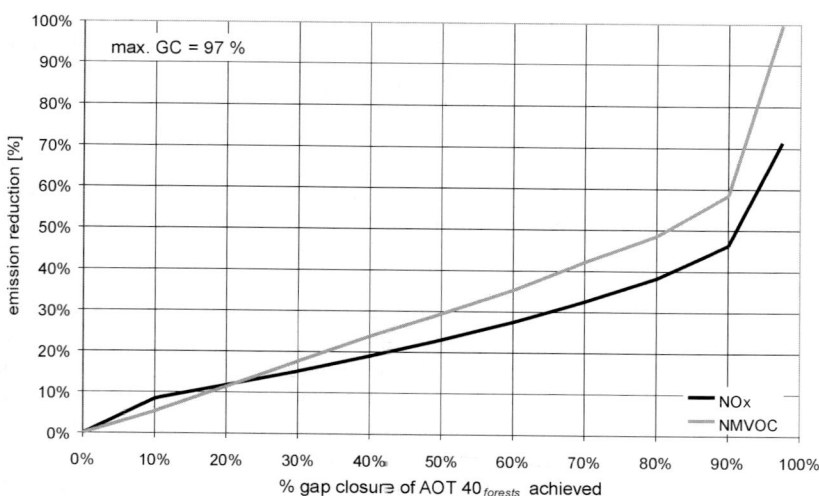

Fig. 11.6. EU total reductions of NO_x and anthropogenic NMVOC to achieve more stringent gap closures of $AOT40_{forests}$; reductions in % on top of trend scenario 2010

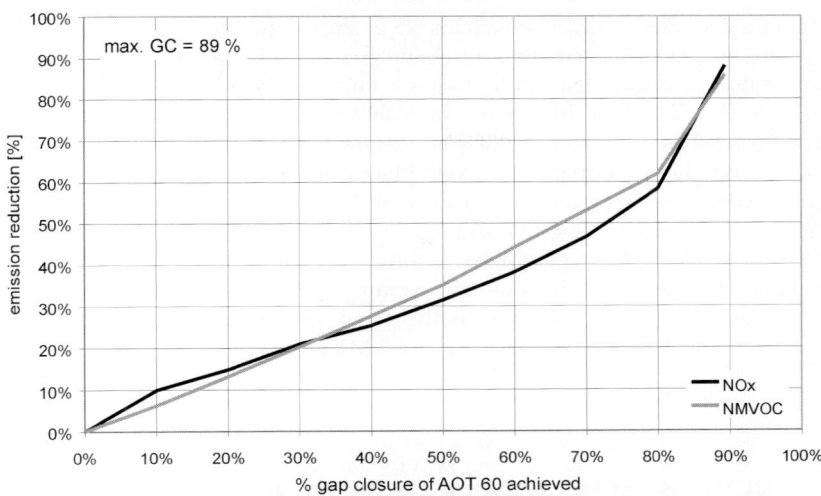

Fig. 11.7. EU total reductions of NO_x and anthropogenic NMVOC to achieve more stringent gap closures of AOT60; reductions in % on top of trend scenario 2010

In addition to that, further source sectors, such as the variety of industrial processes and other mobile sources will have to be addressed within future abatement strategies (the planned directive on emissions from other mobile sources (tractors, construction machinery etc.) would be a step in this direction). In this context, it is useful to find out, to which extent emissions of NO_x and NMVOC would have to be reduced in order to achieve more ambitious gap closures than those analysed in the previous section.

Fig. 11.5 shows the gap closures towards $AOT40_{crops}$ with a threshold of 3 ppm.h wich can be achieved by different emission reductions, estimated with the OMEGA model. As can be seen, even a total abatement of anthropogenic NMVOC, together with a substantial NO_x reduction, in the 15 EU Member States – it is, of course not realistic to assume these reductions to be achieved – would only result in a gap closure of about 68 %. A 50 % gap closure would be achieved by reducing NOx by 55 % and anthropogenic NMVOC emissions by 60 % on top of the trend scenario for 2010. *Figs. 11.6* and *11.7* show emission reductions necessary to achieve different gap closures towards the $AOT40_{forests}$ limit of 10 ppm.h (from April to September) and the AOT60 threshold. For $AOT40_{forests}$, even a 90 % gap closure could be achieved by reducing NOx by 50 % and NMVOC by about 60 %. For AOT60, an 80 % gap closure could be reached by reducing NO_x by 58 % and NMVOC by 62 % on top of the trend scenario for 2010. These results indicate, that at least for $AOT40_{crops}$ and AOT60 it seems to be impossible to achieve a full compliance with the limits, if no exceedances are allowed. The gaps towards $AOT40_{crops}$ are even more difficult to reduce than those for the AOT60 limit. Furthermore, results indicate that for a given gap closure target above ca. 30 %, anthropogenic NMVOC emissions should be reduced to a larger extent than NO_x emissions. However, this is only valid as long as the total emission reductions in all EU countries are examined (as in *Figs. 11.5* to *11.7*). If the results are analysed country by country, a considerable variation of of the emission reductions needed can be found, sometimes between just above 20 % to almost 90 %. *Fig. 11.8* illustrates this, showing the emission reductions per country needed to achieve a 50 % gap closure towards the AOT60 limit. Some countries (Belgium, Germany, the Netherlands and the United Kingdom) would have to reduce anthropogenic emissions of NMVOC far more than their NO_x emissions. At the same time, Austria, Greece and the Scandinavian countries are required to reduce NO_x emissions significantly more than NMVOC. In the remaining countries, as well as on EU average, both ozone precursors have to be reduced by about 32 % (relative to the trend 2010), showing the differences between regions, where NO_x – respectively NMVOC – control is prevalent.

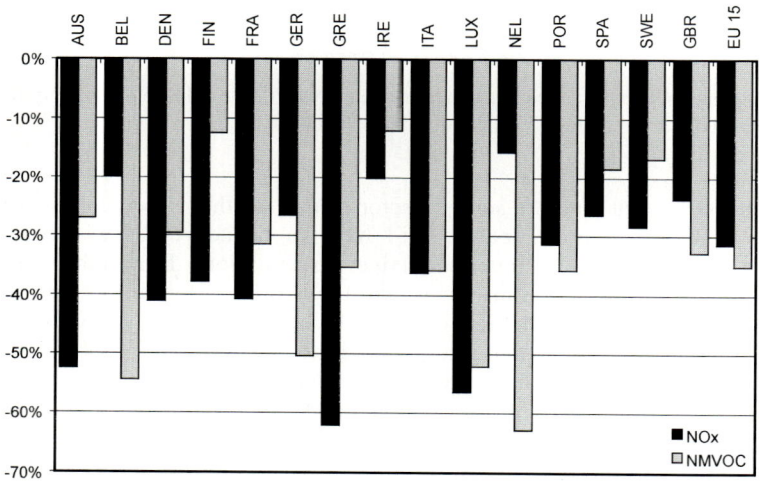

Fig. 11.8. NO_x and NMVOC emission reduction by country to achieve a 50 % gap closure of AOT 60

Modelling results for the three urban areas investigated indicate (*Chapt. 8*), that for reducing local ozone concentrations the reduction of local NMVOC emissions proves to be much more effective than the reduction of local NO_x emissions.

Finally, these results show, that a full compliance with the limits described above in all areas and during all times in Europe is extremely difficult to achieve, if not impossible.

If a gap closure of more than ca. 33% (compared to the 2010 base case situation) is aimed at, the identification of the most efficient strategy (bundle of measures) is difficult, as the costs for the necessary additional abatement measures can only be estimated with increasing uncertainty. However, if actual costs differ from the assumed costs, the efficiency of the strategy will be reduced. One possibility for dealing with this problem is to use market forces to adjust the strategy according to actual abatement costs. This is however much more difficult for ozone than e. g. for greenhouse gases, as the highly non-linear process of ozone formation has to be taken into account. An instrument that is capable of doing this is the newly developed 'iterative zonal permit trading'. Under this approach emissions budgets (or targets) are set for each country and each precursor according to the estimated optimal strategy. Within each country, simple "one-for-one" emissions trading is then used to implement the targets, but no trading is allowed between countries. However, on a periodic basis (e.g. every 3-5 years) the information provided by the permit prices in each country is used to adjust the emissions budgets in an iterative process. In comparison with alternative "economic" implementation mechanisms, iterative zonal permit trading has a number of important advantages: it can incorporate a large number of receptor points; it ensures that ambient air quality targets are never violated at any of these points; it provides information about the cost impact on each country; it is relatively simple to implement; and it represents only an incremental change to the existing institutional framework for international environmental agreements. A main conclusion is, that due to the uncertainties involved, strategies to reduce air pollution should not be seen as irrevocable after implementation, but should be re-assessed, negotiated and adjusted after some years have passed.

Costs and benefits of the implementation of abatement strategies are not equally distributed among EU countries. Some national economies have to cope with a greater burden of imposed additional costs than others. In addition, the benefits gained from the reduced ozone concentrations is not proportional to the costs in the different countries. An example is Ireland, where more benefits from the emission reduction occur in the UK than in Ireland itself. Thus, to achieve a fair and equitable distribution, burden sharing rules have been investigated and applied, using the following principles:

- Polluter Pays Principle   (PPP)
- Victim Pays Principle   (VPP)
- Ability to Pay Principle   (APP)

In addition to these, a combined approach was used, called Balancing Competing Principles (BCP) and using a combination of the principles listed above (PPP50% : VPP5% : APP45%). If the costs a country has to bear by implementing the emission reduction measures is below the amount allocated to the country by one of this principles, the country should pay the difference into a so-called ozone fund. If the costs exceed the allocated amount, the difference is paid to the country from the ozone fund. Under PPP, each country has to pay a share relative to the share of total damage caused by it, calculated by assessing its contribution to total ozone formation. This results in additional payments especially by Germany and Belgium. Applying VPP, countries pay according to their share of the benefits, calculated by monetarising the avoided damage, e.g. avoided crop losses and health impacts, using the willingness-to-pay approach as developed within the ExternE projects supported by DGXII. With VPP, main payers are Italy and Greece followed by Spain and France, whereas Germany is receiving money. APP uses the Gross Domestic Product as a way to measure the relative economic ability of each country to cope with additional costs; countries like Germany, UK and the Netherlands would have to make additional payments. With BCP, the largest amount would have to be paid by Germany, followed by the UK, the Netherlands and Belgium, whereas the largest transfer payemtns would go to Spain.

Table 11.6. Transfer payments under different burden sharing rules

Country	Transfer payments (in bill. EURO) under			
	APP	PPP	VPP	BCP
Austria	0.42	0.03	-0.34	0.19
Belgium	0.52	0.45	-0.35	0.44
Denmark	0.06	-0.24	-0.56	-0.12
Finland	-0.34	-0.41	-0.84	-0.40
France	-1.03	-0.35	0.51	-0.61
Germany	5.29	2.09	-3.24	3.27
Greece	-0.60	0.23	2.81	-0.01
Ireland	-0.29	-0.26	-0.58	-0.29
Italy	-2.61	0.14	5.50	-0.83
Luxembourg	0.01	0.05	-0.04	0.03
Netherlands	1.09	0.09	-0.41	0.52
Portugal	-0.48	0.10	-0.14	-0.17
Spain	-3.86	-1.55	0.86	-2.47
Sweden	-0.16	-0.40	-1.03	-0.32
United Kingdom	1.97	0.02	-2.17	0.79

Note: + means *payment* into the fund, – means *receiving* from the fund

There are countries, that would benefit from all four principles, especially Ireland, Sweden and Finland; Germany and to less extent the United Kingdom would benefit only by the VPP principle, whereas Spain and France profit from all principles except VPP (see *Table 11.6*).

Since none of the three principles can be considered superior on any objective basis, it is not clear which should be adopted. However, in terms of finding a solution that EU member States could agree upon, BCP might be a possibility, since it incorporates the contribution to the problem as well as the ability of the country to handle the additional payment without putting too much strain on its budget.

Conclusions

A modelling framework consisting of coupled models dealing with emissions, atmospheric transport and chemical transformation, environmental, health and macroeconomic impacts, that is able to identify efficient strategies to reduce tropospheric ozone concentrations in Europe and assess the environmental and economic impacts has been developed. The model system was tested and its use demonstrated by analysing scenarios with different emissions of ozone precursor substances for Europe as a whole and selected European cities.

The results clearly indicate that the implementation of policies and legislation in place and in pipeline will lead to a significant reduction of ground-level ozone concentrations all over Europe. However, considerable exceedances of ozone limit values as described in the current proposal of the European Commission for the ozone Directive will still occur in the year 2010 and beyond. Additional action has to be taken to reduce the emissions of anthropogenic precursors of ozone, NO_x and NMVOC, even more than projected for the base case (trend) scenario.

Outlook

The methods and models applied in this study present a proficient tool for the identification and evaluation of strategies for the efficient abatement of tropospheric ozone in Europe. By including acidification into the optimisation process, a step into the direction of an integrated environmental assessment has already been taken. Both, the methodology developed and the model framework used, can be extended to incorporate various pollutants and effects (e.g. ozone, acidification, eutrophication, global warming and urban air quality), thus allowing for a harmonised and balanced assessment of costs and benefits, as well as macroeconomic and distributional effects of strategies to improve air quality over Europe.

Index

a

abatement 7, 10, 14, 20, 25, 35, 37, 39, 41, 43, 53, 56, 61, 62, 75, 76, 77, 83, 93, 96, 97, 99, 102, 115, 120, 148, 150, 151, 161 166, 170, 176, 189, 195, 203, 205, 212, 213
  costs 2, 3, 36, 38, 46, 47, 48, 50, 51, 55, 58, 66, 69, 100, 104, 105, 106, 107, 108, 109, 111, 112, 113, 114, 119, 152, 155, 156, 158, 159, 160, 162, 163, 164, 165, 167, 168, 169, 171, 173, 174, 175, 177, 178, 179, 180, 181, 182, 186, 188, 193, 196, 197, 198, 200, 202, 208, 209, 214, 215, 217
  measures 46, 48, 51, 54, 55, 101, 104, 208, 216
air quality 14, 16, 17, 18, 68, 93, 96, 97, 121, 138, 143, 144, 148, 149, 150, 153, 159, 167, 172, 176, 205, 215, 217
ambient 1, 11, 17, 18, 101, 122, 149, 153, 154, 155, 156, 157, 159, 160, 161, 167, 182, 205, 207, 215
anthropogenic 5, 10, 21, 27, 28, 30, 31, 84, 92, 115, 116, 176, 205, 208, 213, 214, 217
assessment 7, 14, 17, 18, 46, 58, 61, 62, 70, 99, 107, 109, 112, 119, 120, 121, 122, 123, 128, 131, 138, 174, 184, 217
atmosphere 1, 14, 26, 36, 127

b

benefit 58, 101, 140, 148, 165, 168, 172, 176, 201, 216
boundary layer 1, 96, 128, 137, 211

c

coating 52, 53, 55, 70, 71
cold start 37, 38, 41, 42, 212

combustion 5, 7, 15, 25, 26, 28, 29, 35, 36, 41, 42, 43, 44, 46, 47, 48, 50, 51, 56, 65, 69, 72, 89, 104, 124
concentration 1, 3, 11, 13, 16, 19, 70, 83, 99, 101, 112, 121, 122, 127, 128, 129, 135, 136, 137, 138, 139, 144, 147, 148, 153, 155, 182, 184, 207, 208
critical 87, 93
  levels 12, 13, 16, 23, 107
  loads 15, 99
crop
  damage 114, 184
  yield 1, 10, 11, 14, 114

d

directive 17, 19, 47, 48, 51, 52, 56, 74, 128, 148, 212
dispersion 20, 33, 120, 124, 127, 136, 144

e

ecosystems 10, 12, 13, 14, 18, 107, 112
emissions 1, 2, 3, 36, 38, 41, 42, 43, 44, 52, 61, 62, 63, 64, 66, 67, 81, 83, 86, 87, 88, 90, 93, 97, 104, 109, 114, 118, 121, 122, 123, 124, 133, 135, 142, 149, 150, 151, 153, 154, 156, 157, 158, 159, 160, 161, 162, 163, 164, 165, 166, 167, 168, 173, 174, 175, 181, 182, 185, 186, 193, 194, 195, 196, 205, 208, 209, 215, 216, 217
  biogenic 5, 8, 10, 21, 27, 80, 84, 89, 108, 125, 138
  NMVOC 8, 9, 10, 27, 28, 29, 30, 32, 35, 46, 51, 52, 68, 70, 71, 73, 74, 75, 76, 77, 80, 100, 111, 112, 115, 116, 119, 126, 206, 213, 214

Index

NO_x 7, 8, 9, 10, 13, 14, 15, 21, 25, 26, 27, 29, 30, 31, 32, 42, 43, 46, 51, 58, 68, 69, 71, 72, 73, 76, 77, 78, 79, 89, 92, 99, 103, 107, 108, 110, 111, 112, 115, 127, 128, 131, 132, 134, 139, 140, 141, 143, 145, 146, 147, 148, 155, 169, 171, 183, 184, 206, 212, 214

employment 193, 196, 197, 198, 200, 201, 202, 209

exceedance 1, 3, 19, 100, 107, 115, 118, 128, 130, 131, 132, 133, 135, 138, 139, 140, 141, 142, 144, 206, 210, 211

f

fuel cell 37, 44, 46

g

good housekeeping 53

ground-level ozone 10, 11, 14, 15, 19, 21, 35, 58, 59, 99, 110, 114, 119, 205, 208, 209, 211, 217

h

health 1, 10, 11, 13, 14, 16, 17, 18, 19, 20, 22, 43, 86, 93, 107, 111, 112, 113, 114, 115, 121, 184, 185, 195, 205, 209, 216, 217

hydrocarbons 5, 7, 16, 21, 43, 44, 96, 97, 120

i

impacts 2, 3, 10, 11, 13, 16, 22, 61, 114, 171, 174, 181, 185, 193, 195, 201, 202, 208, 209, 216, 217

integrated 14, 16, 17, 48, 99, 107, 120, 122, 124, 174, 184, 217

inventory 7, 32, 33, 46, 51, 79, 80, 93, 123, 124, 125, 126, 128, 137, 150

l

large combustion plants (LCP) 26, 47, 49, 58, 69, 72

legislation 2, 41, 56, 61, 67, 68, 69, 70, 76, 125, 205, 211, 217

light-off 41

limit value 1, 2, 3, 10, 13, 15, 16, 18, 66, 69, 128, 206, 207, 208, 209, 217

m

meteorology 86, 88, 92, 175

mobile sources 5, 7, 25, 26, 28, 29, 76, 78, 89, 212

n

non-linear 1, 13, 89, 92, 93, 100, 102, 108, 116, 119, 153, 154, 155, 159, 163, 182, 209, 215

o

objectives 17, 19, 20, 152, 153, 154, 156, 158

ozone formation 1, 5, 6, 7, 10, 13, 20, 21, 83, 87, 93, 97, 120, 121, 146, 147, 215, 216

p

paint 28, 63

particulate 42, 43, 44

peak 6, 11, 13, 14, 19, 20, 42, 101

per capita 30, 72, 178, 179, 180

pollutant 1, 7, 13, 16, 17, 18, 43, 68, 75, 100, 101, 103, 104, 115, 127, 128, 136, 137, 138, 144, 148, 153, 208

power generation 15, 26, 27, 29, 47, 48, 56, 71, 78

precursor 1, 2, 7, 25, 36, 61, 64, 66, 83, 114, 121, 135, 153, 154, 155, 157, 161, 166, 174, 183, 205, 208, 215, 217

primary measures 47, 48, 69, 71, 72, 212

printing 28, 52, 70, 135

processes 1, 7, 15, 20, 26, 27, 28, 41, 43, 44, 47, 48, 51, 52, 53, 54, 63, 66, 70, 72, 73, 78, 89, 93, 150, 193, 194, 196, 202, 212

r

respiratory 1, 11, 22

road transport 5, 7, 8, 15, 26, 27, 30, 35, 36, 37, 45, 46, 51, 56, 62, 66, 67, 68, 76, 77, 78, 89, 90, 93, 104, 124, 205

rubber 1, 52, 70, 185

s

secondary measures 47, 49

solvent use 8, 28, 35, 36, 51, 53, 56, 63, 69, 70, 74, 75, 78

stationary sources 5, 53, 72, 125

t

target value 1, 19, 20, 155
three–way catalyst 37
threshold 13, 16, 18, 19, 20, 22, 86, 87, 99, 101, 107, 109, 110, 121, 122, 128, 129, 131, 133, 134, 137, 138, 139, 143, 144, 148, 205, 207, 208
transboundary 1, 14, 15, 16, 20, 119, 125, 151, 159, 161, 173, 175
transport 1, 5, 7, 8, 14, 15, 16, 20, 25, 26, 27, 28, 104, 125, 127, 135, 217
troposphere 1, 5

u

uncertainty 86, 93, 116, 165, 173, 215

v

volatile organic compounds 1, 5, 15, 96, 100, 149

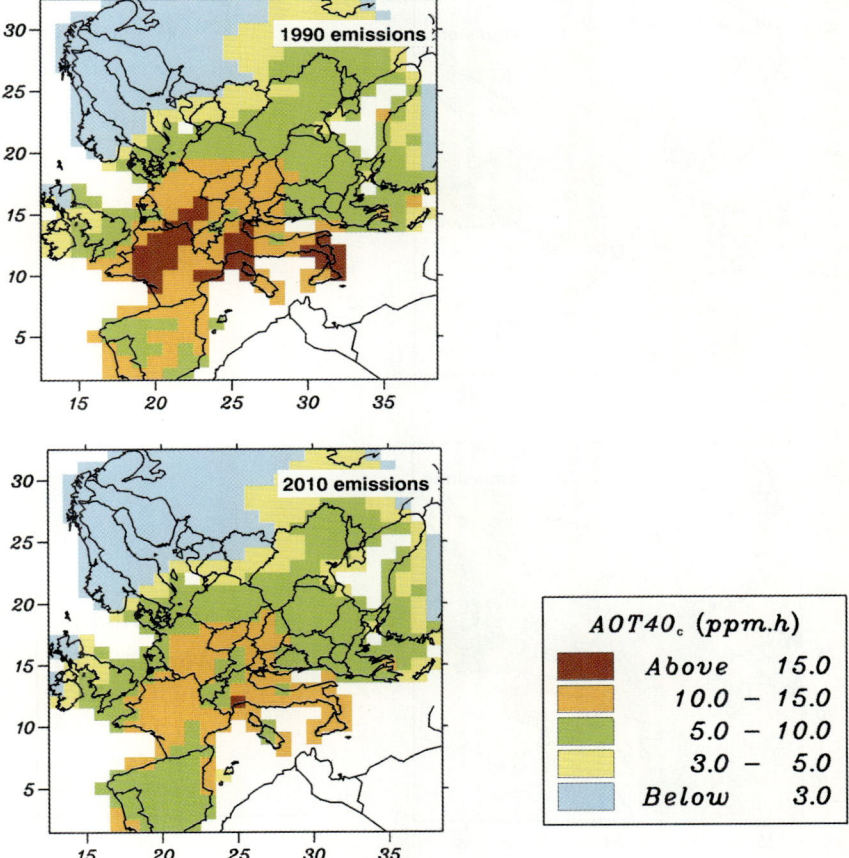

Plate 1. Predicted 3-monthly AOT40$_{crops}$ for 1990 and Trend Scenario 2010 emissions. Calculations with EMEP MSC-W Lagrangian model with 5-years of meteorology.

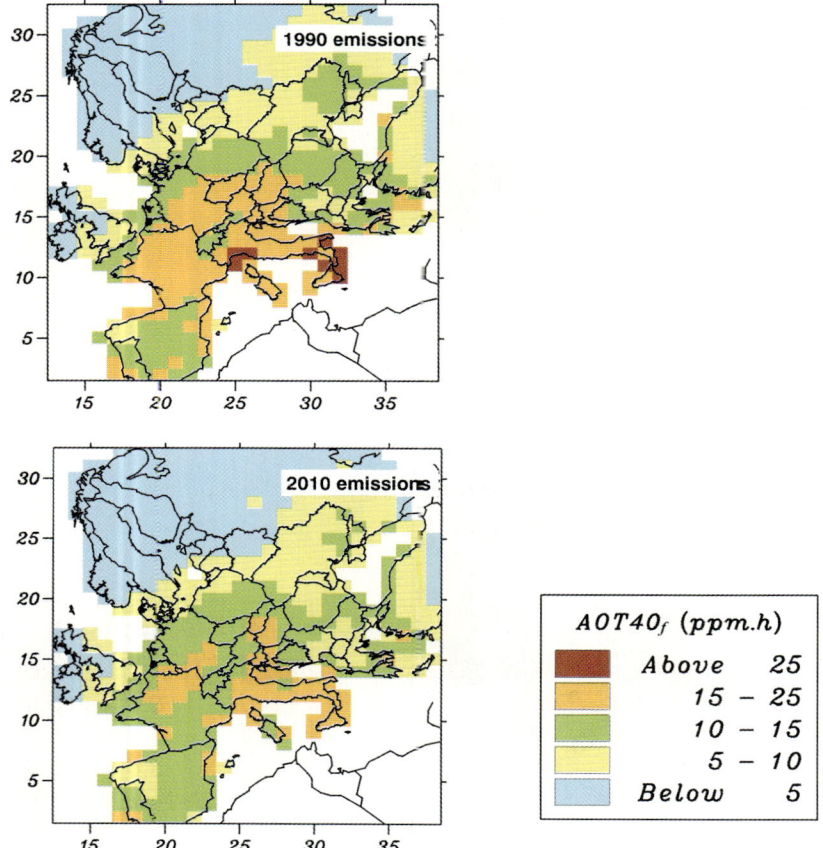

Plate 2. Predicted 6-monthly AOT40$_{forests}$ for 1990 and Trend Scenario 2010 emissions. Calculations with EMEP MSC-W Lagrangian model with 5-years of meteorology.

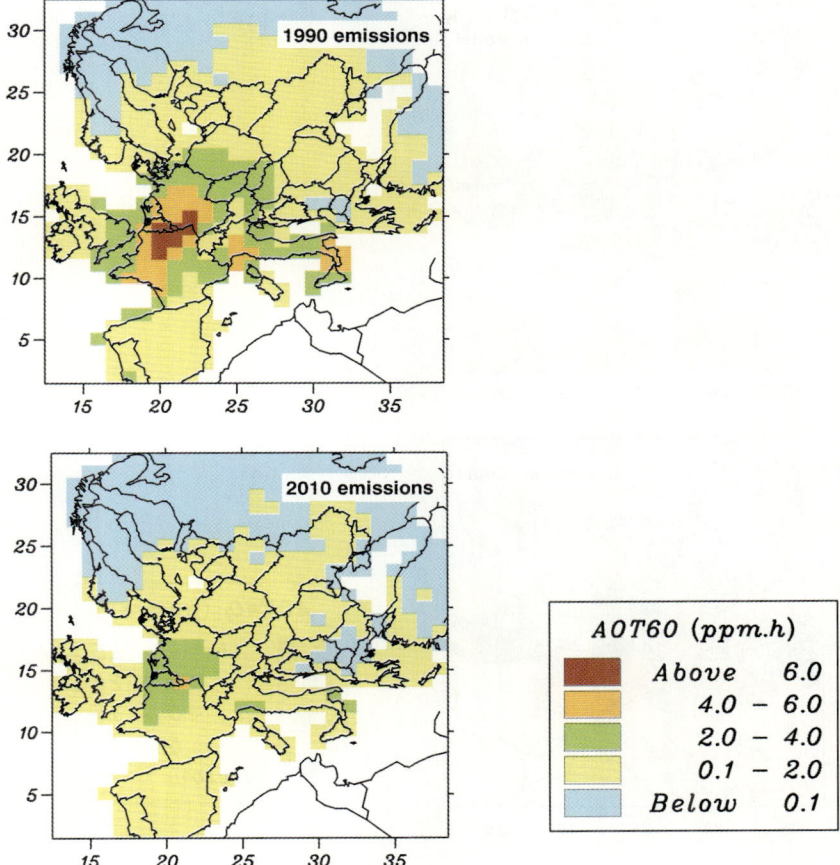

Plate 3. Predicted 6-monthly AOT60 (health) for 1990 and Trend Scenario 2010 emissions. Calculations with EMEP MSC-W Lagrangian model with 5-years of meteorology.

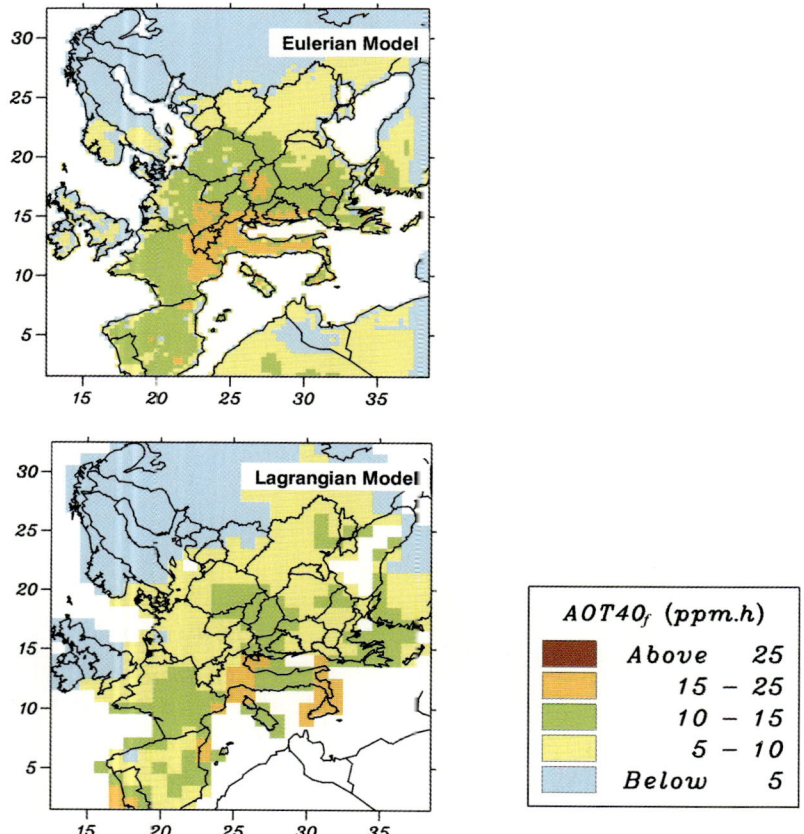

Plate 4. Comparison of Eulerian (top) and Lagrangian (bottom) model predictions of 6-monthly AOT40$_{forests}$. Emissions from Trend Scenario 2010, meteorology of April – September 1996.

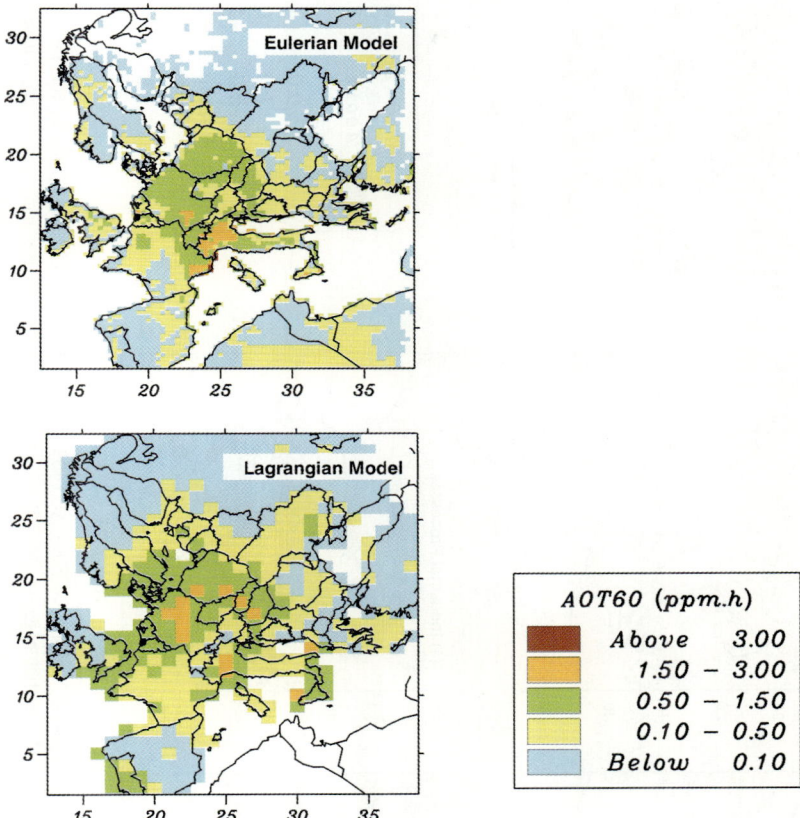

Plate 5. Comparison of Eulerian (top) and Lagrangian (bottom) model predictions of 6-monthly AOT60 (health). Emissions from Trend Scenario 2010, meteorology of April – September 1996.

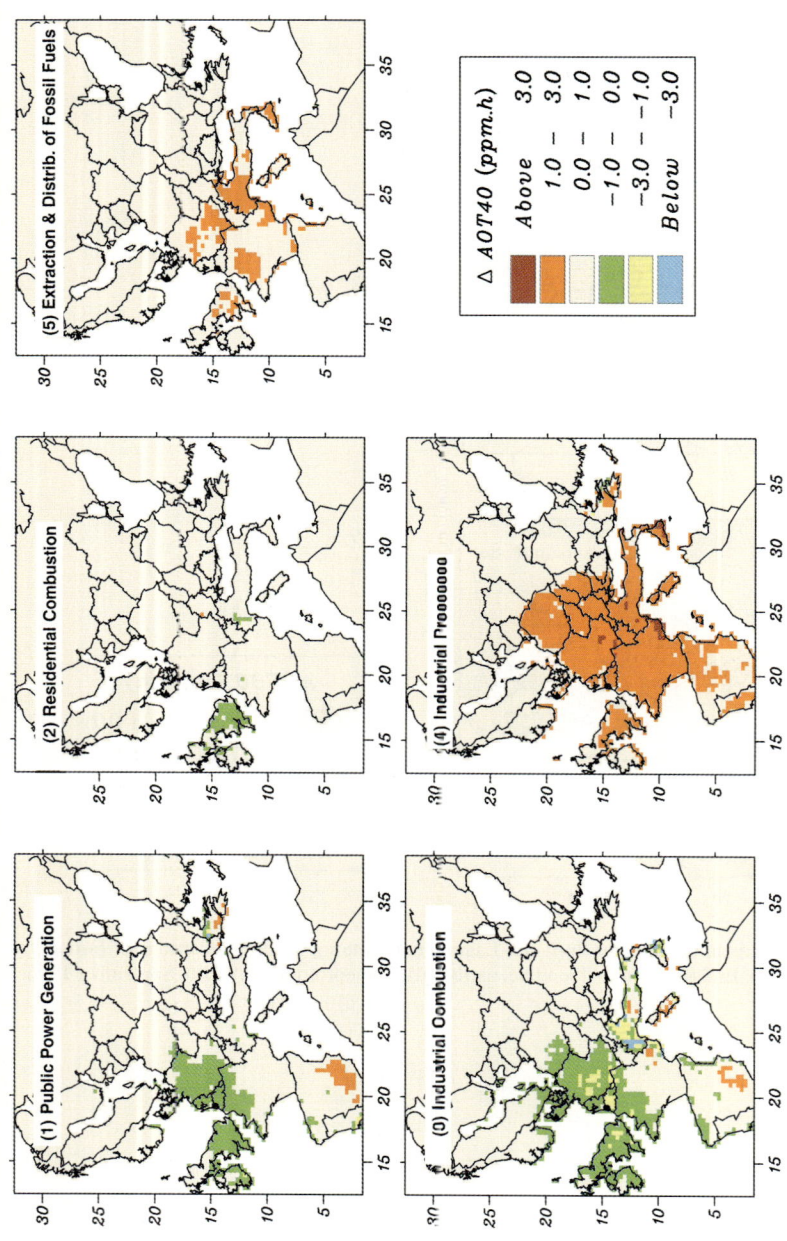

Plate 6. Effect of reducing emissions from SNAP sectors 1 – 5 on AOT40$_{forests}$. Calculations with Eulerian model for Trend Scenario 2010 emissions, meteorology of April – September 1996. (Note that negative AOT40$_{forests}$ means that AOT40 has increased when emissions from sector removed.)

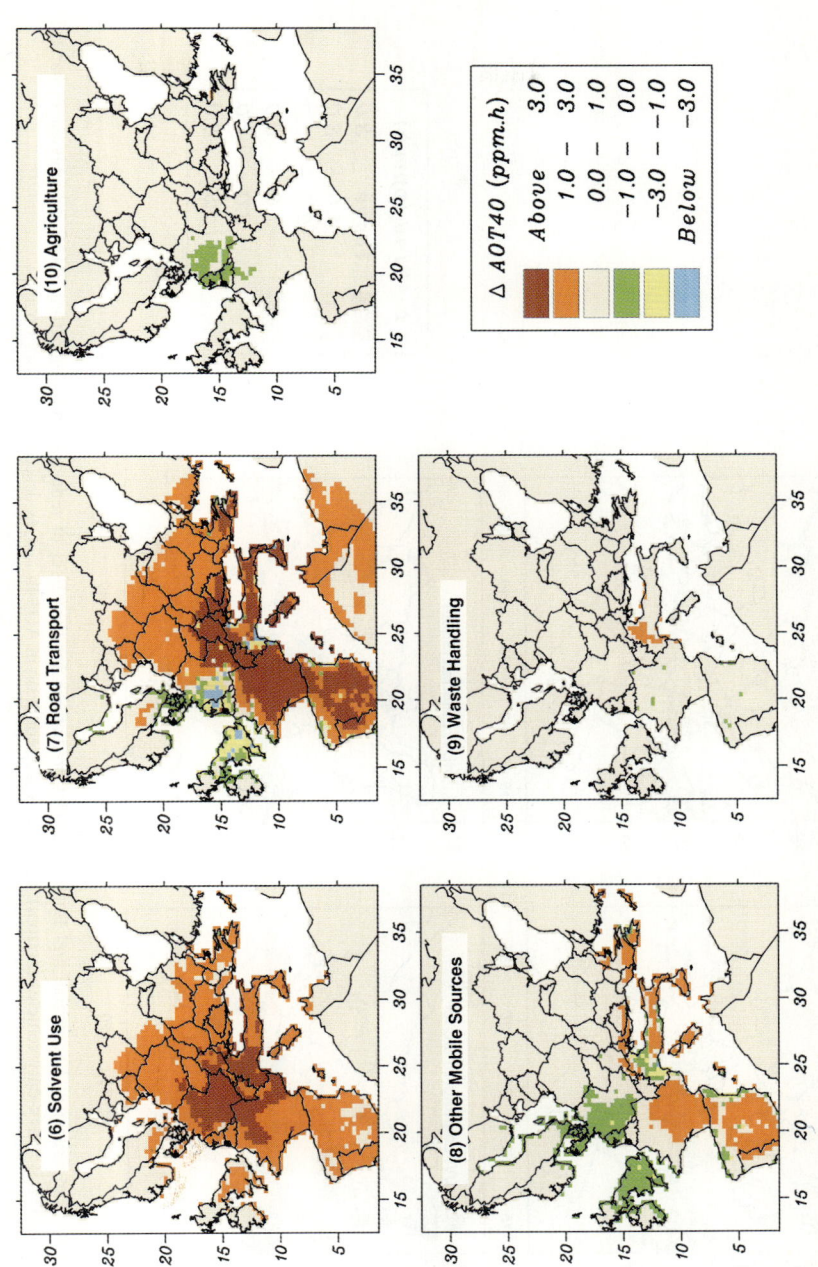

Plate 7. Effect of reducing emissions from SNAP sectors 6 – 10 on AOT40$_{forests}$. Calculations with Eulerian model for Trend Scenario 2010 emissions, meteorology of April – September 1996. (Note that negative AOT40$_{forests}$ means that AOT40 has increased when emissions from sector removed.)

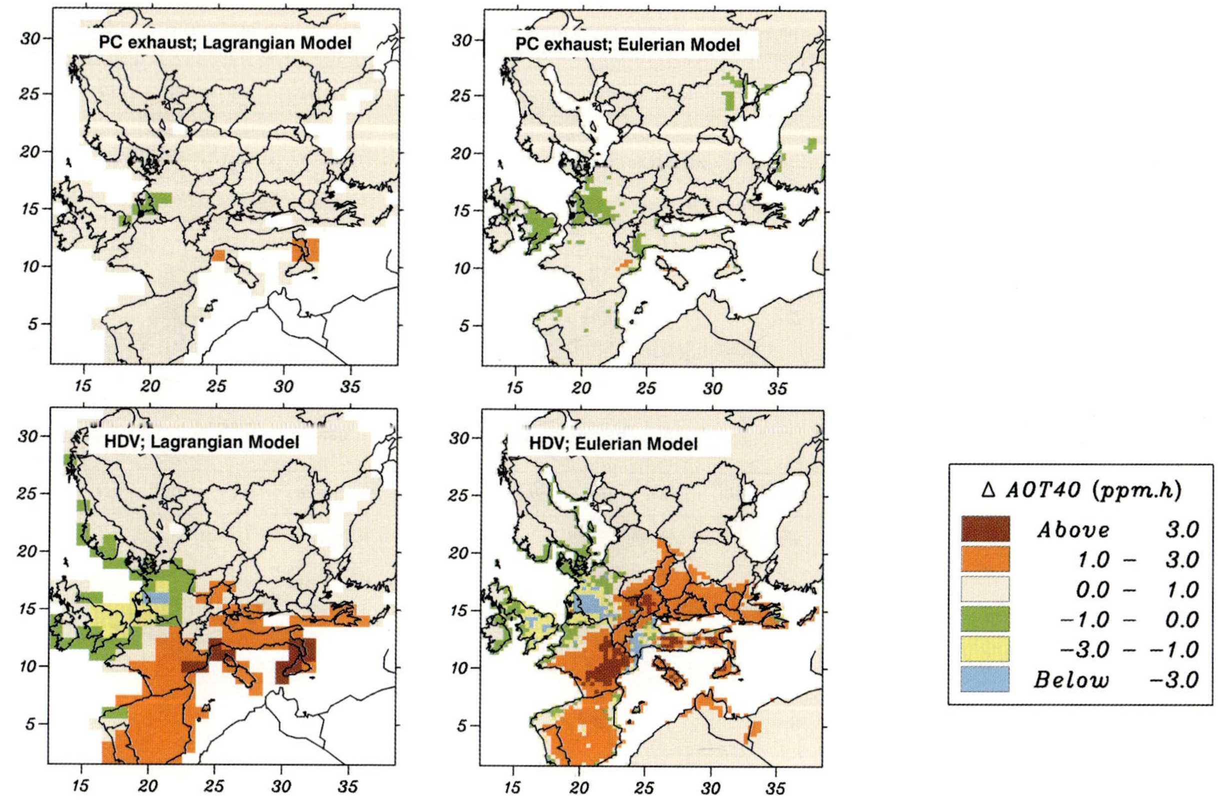

Plate 8. Comparison of Lagrangian and Eulerian model predictions for the effect of reducing emissions from SNAP 7.1: passenger car exhaust (top) and SNAP 7.3: heavy-duty vehicles (bottom). Plots show the reduction in AOT40$_{forests}$, as described for *Plates 6* and *7*

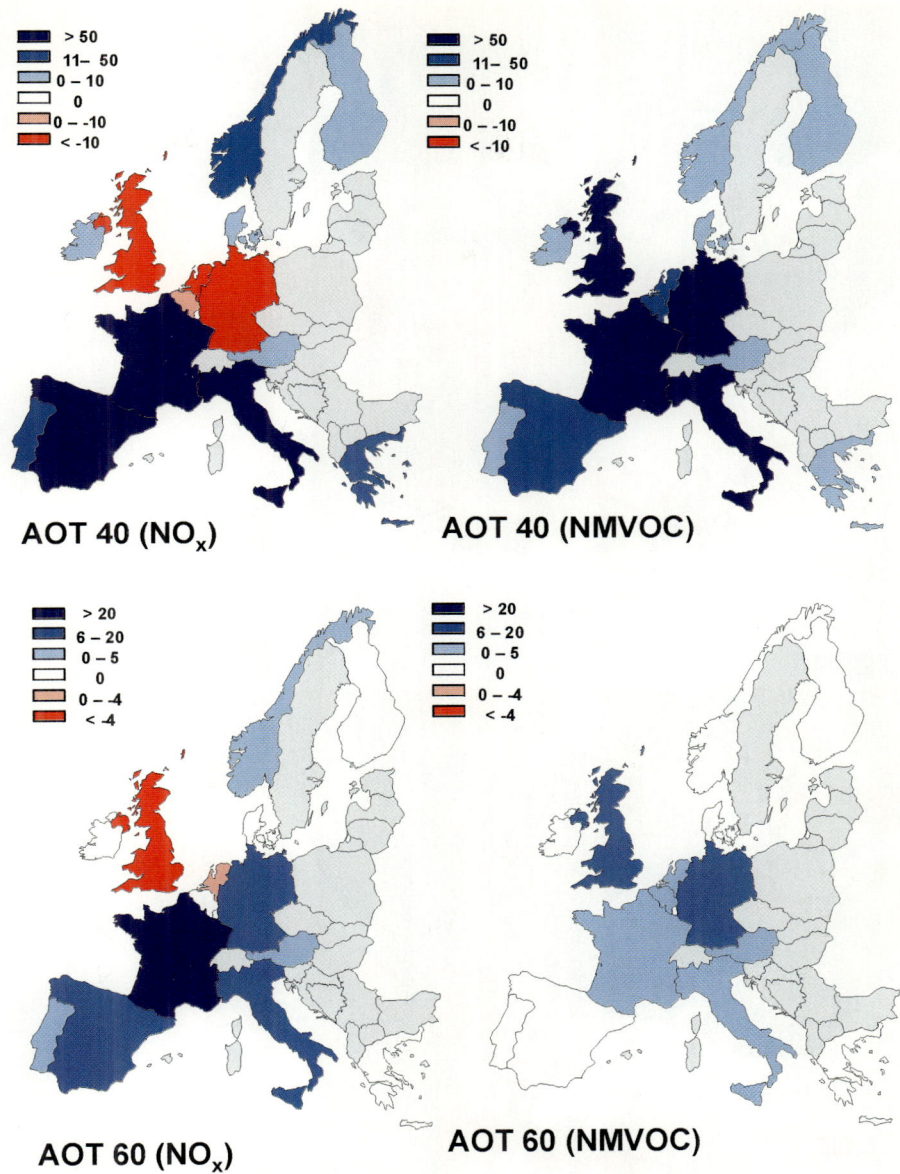

Plate 9: Calculated contribution of each countries NO$_x$ and NMVOC emissions to reducing average European AOT40$_{crops}$ and AOT60. Numbers are in ppb.h and are derived from 40 % emission reduction scenarios. (see *Sect. 6.7 Tables. 6.6* to *6.9*)

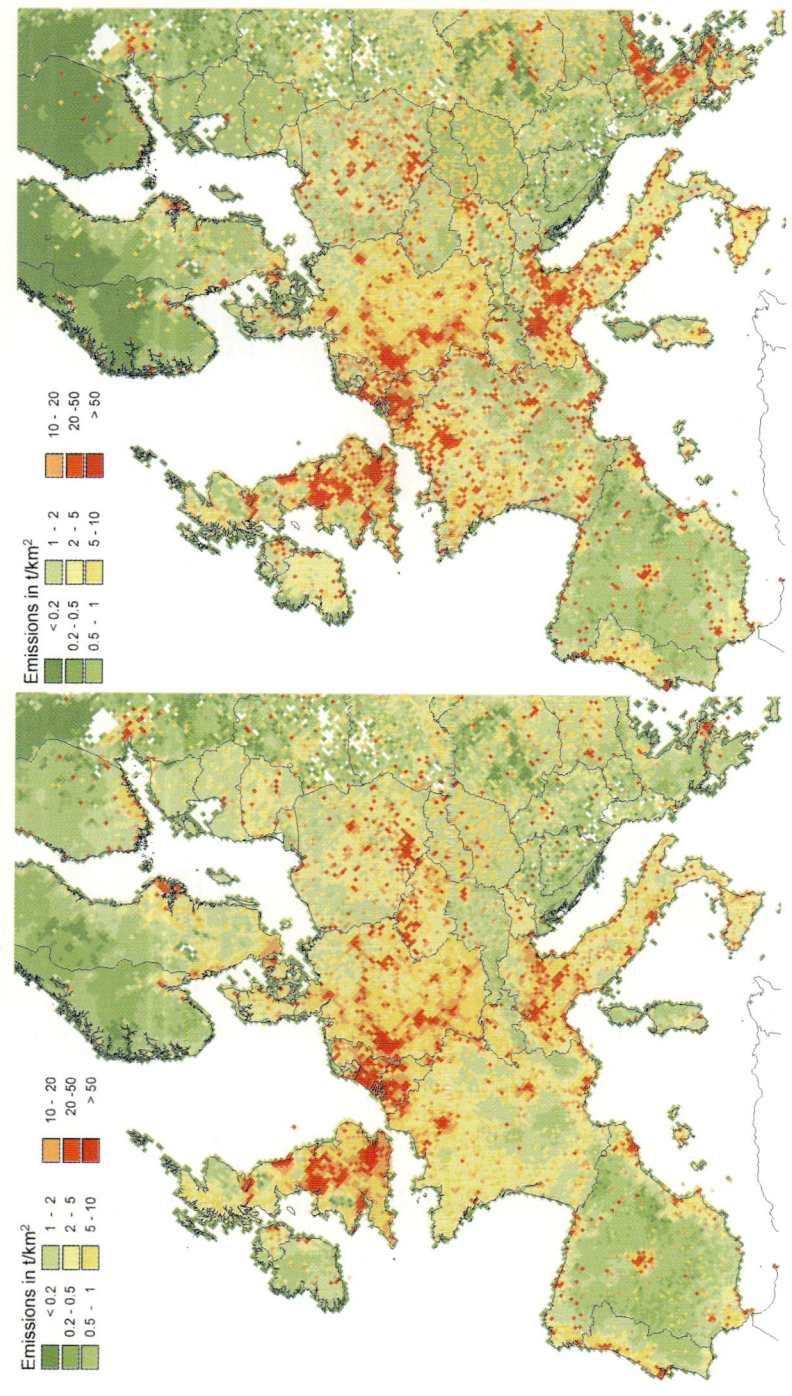

Plate 10. Anthropogenic emissions of NO$_x$ (left) and NMVOC (right) in Europe at a spatial resolution of 20 × 20 km in the year 1994 *(maps by U. Schwarz, IER)*

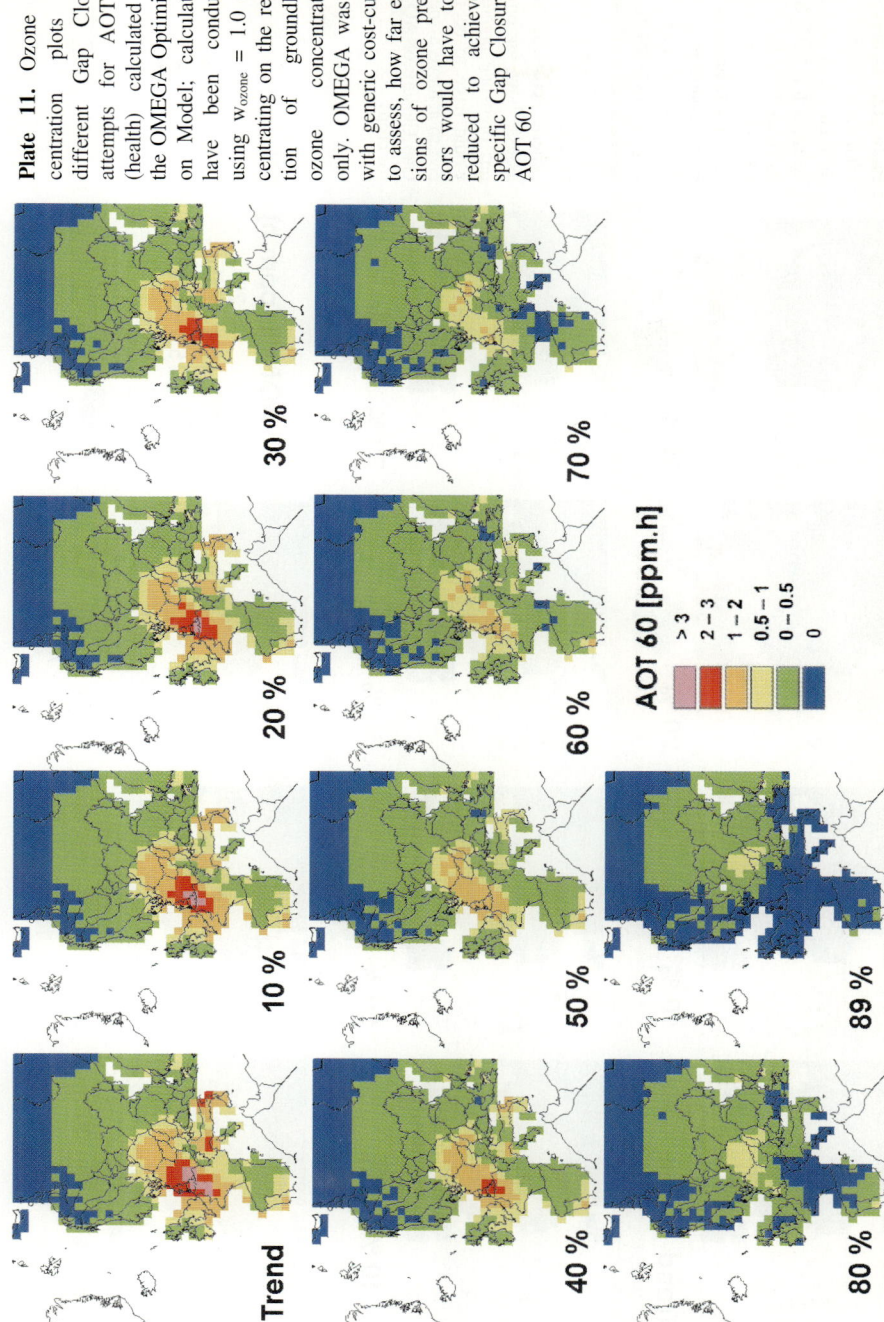

Plate 11. Ozone concentration plots for different Gap Closure attempts for AOT 60 (health) calculated by the OMEGA Optimisation Model; calculations have been conducted using $w_{ozone} = 1.0$ concentrating on the reduction of groundlevel ozone concentrations only. OMEGA was run with generic cost-curves to assess, how far emissions of ozone precursors would have to be reduced to achieve a specific Gap Closure of AOT 60.

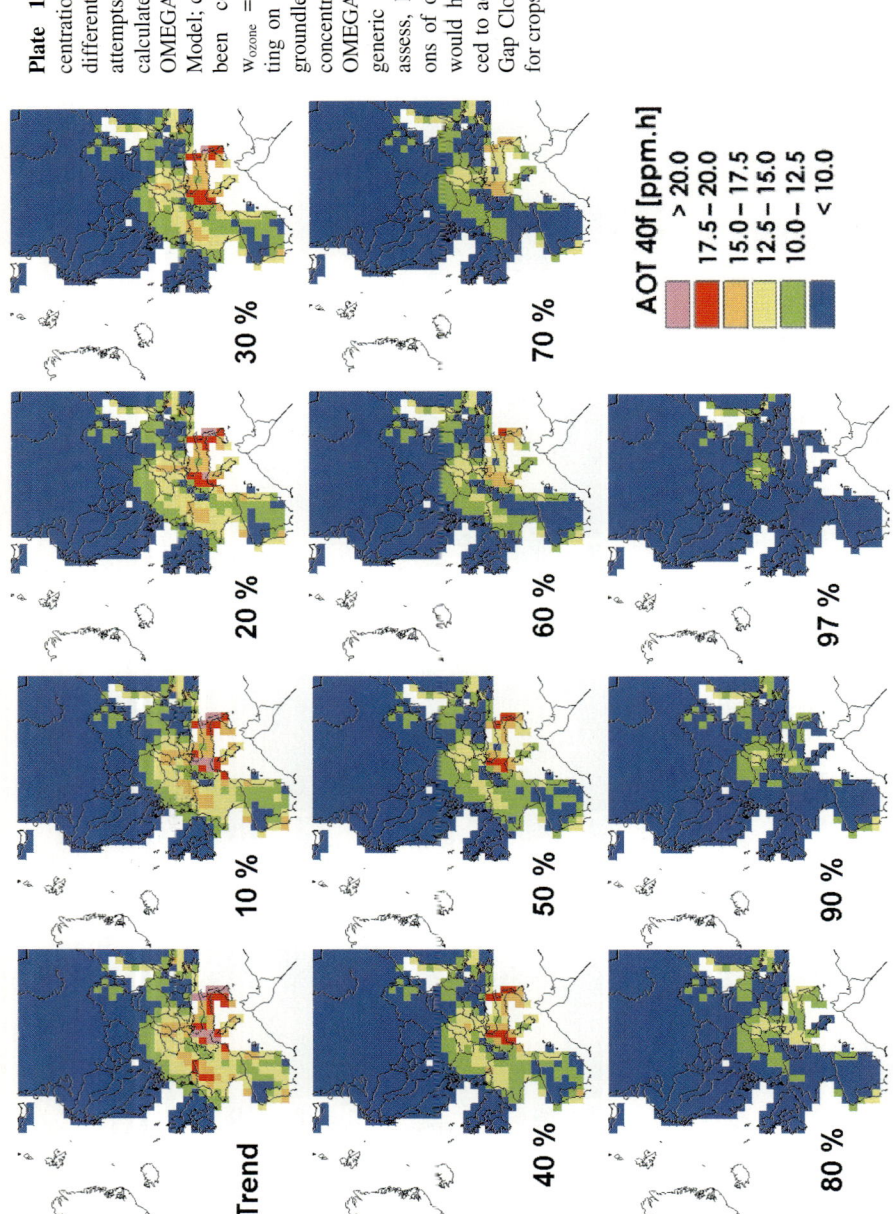

Plate 12. Ozone concentration plots for different Gap Closure attempts for AOT 40$_{crops}$ calculated by the OMEGA Optimisation Model; calculations have been conducted using $w_{ozone} = 1.0$ concentrating on the reduction of groundlevel ozone concentrations only. OMEGA was run with generic cost-curves to assess, how far emissions of ozone precursors would have to be reduced to achieve a specific Gap Closure of AOT 40 for crops

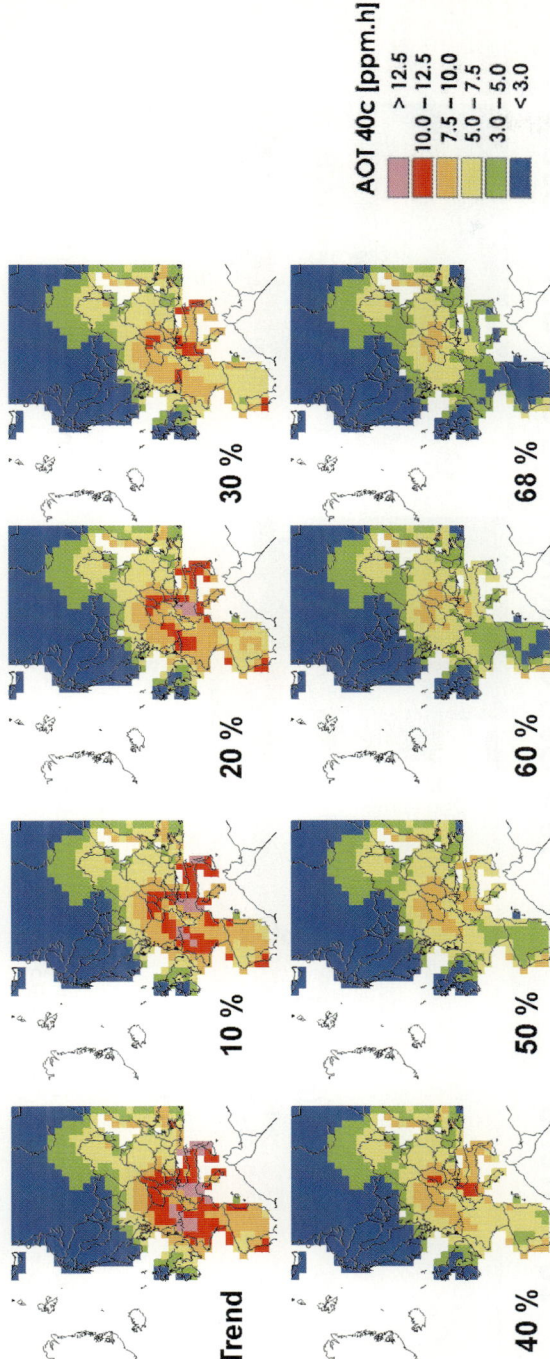

Plate 13. Ozone concentration plots for different Gap Closure attempts for AOT 40$_{forests}$ procedure as described for *Plate 11* and *12*

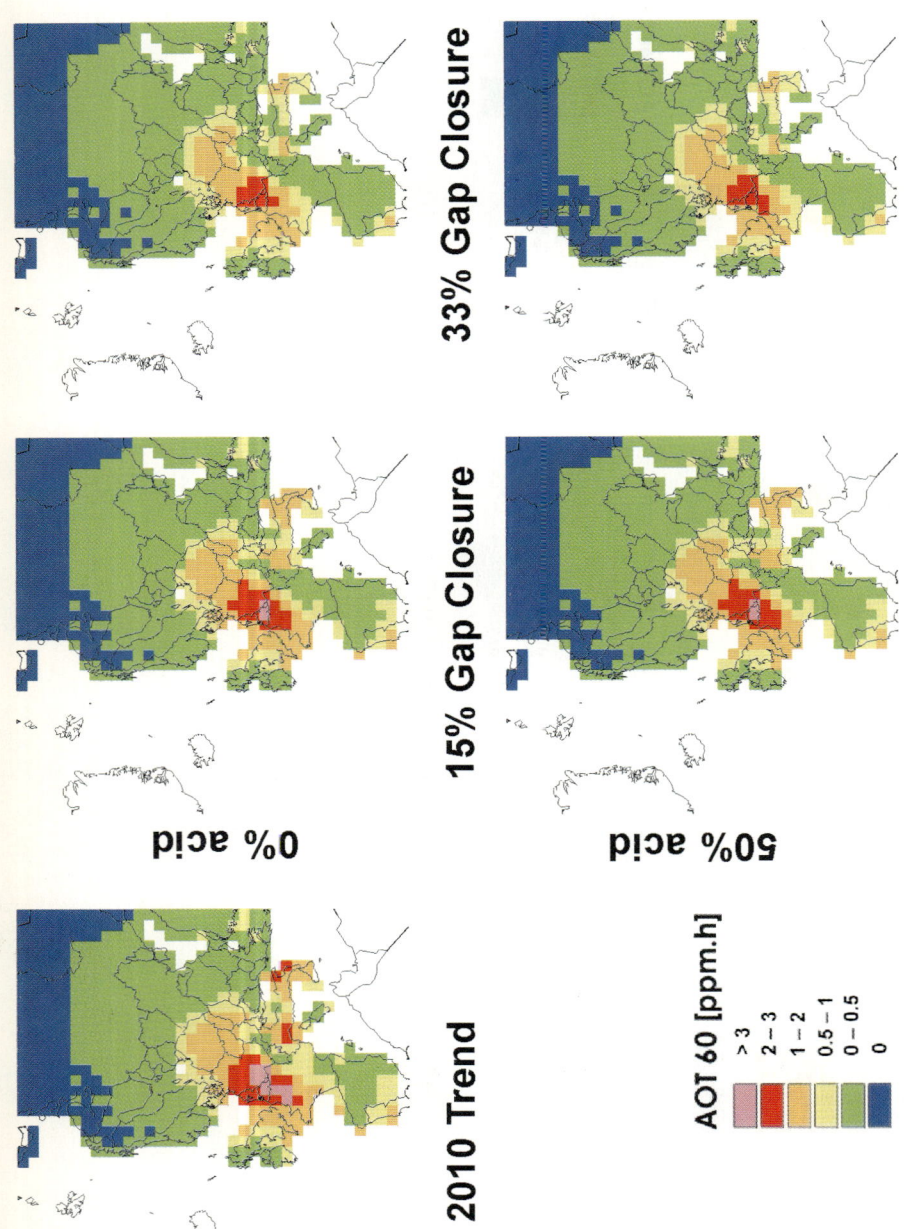

Plate 14. Ozone concentration plots for an optimisation scenario for AOT 60 (health) calculated by the OMEGA Optimisation Model using $w_{ozone} = 1.0$ (top) and $w_{ozone} = w_{acid} = 0.5$ (bottom)

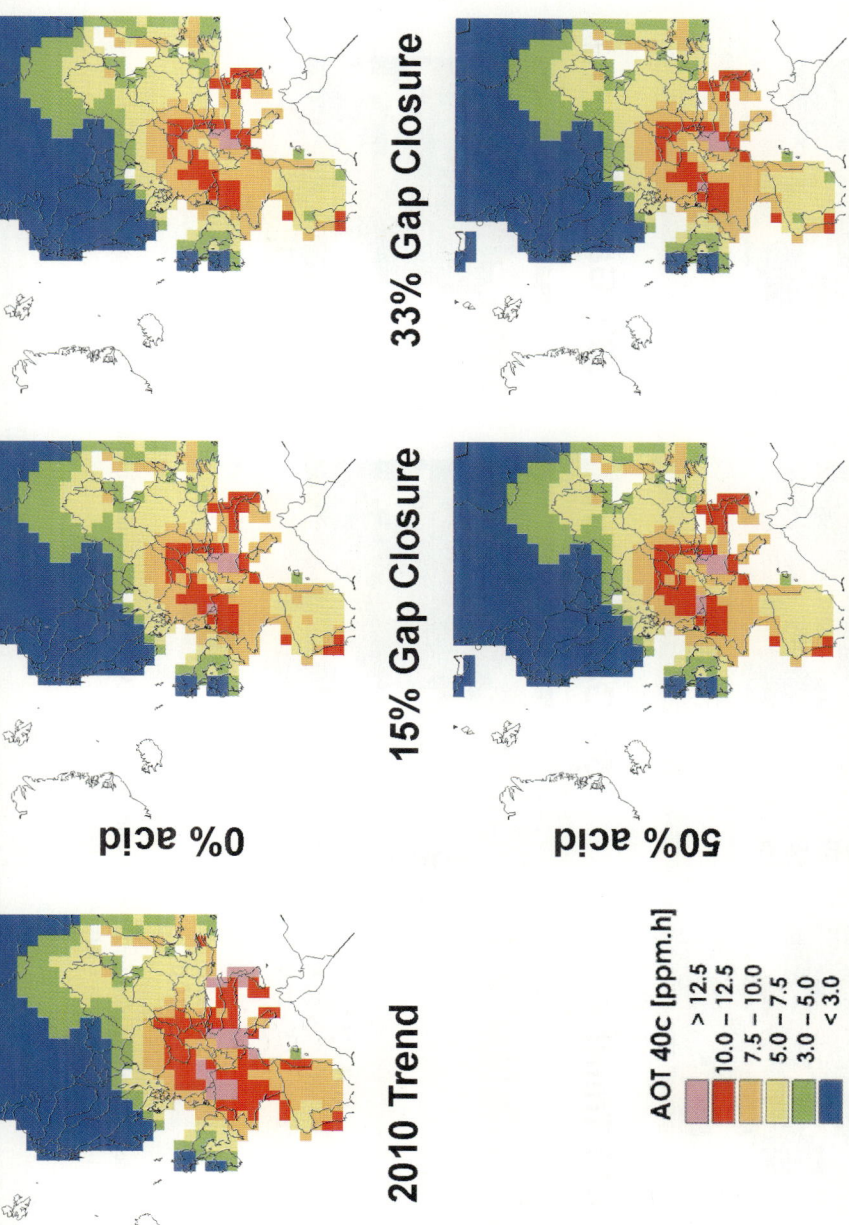

Plate 15. Ozone concentration plots for an optimisation scenario for AOT 40_{crops} calculated by the OMEGA Optimisation Model using $w_{ozone} = 1.0$ (top) and $w_{ozone} = w_{acid} = 0.5$ (bottom)

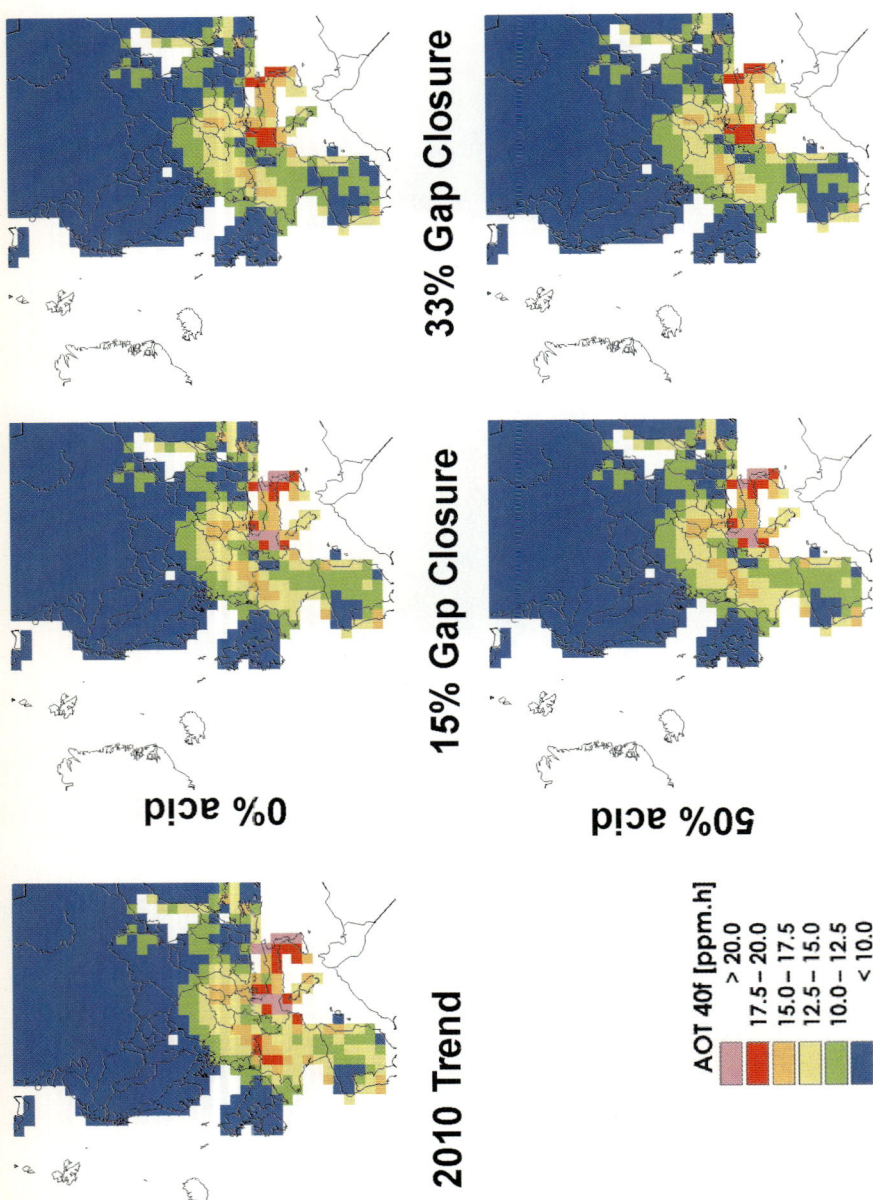

Plate 16. Ozone concentration plots for an optimisation scenario for AOT $40_{forests}$ calculated by the OMEGA Optimisation Model using $w_{ozone} = 1.0$ (top) and $w_{ozone} = w_{acid} = 0.5$ (bottom)

Printing: Mercedes-Druck, Berlin
Binding: Stürtz AG, Würzburg